Price Guide to
COLLECTIBLE
KITCHEN
APPLIANCES

Gary Miller and K.M. Scotty Mitchell

Price Guide to COLLECTIBLE KITCHEN APPLIANCES

■

Wallace-Homestead
Book Company
Radnor, Pennsylvania

Published in Radnor, Pennsylvania 19089, by Wallace-Homestead,
a division of Chilton Book Company

Designed by Tracy Baldwin
Manufactured in the United States of America

Library of Congress Cataloging in Publication Data

Miller, Gary, 1940–
     Price guide to collectible kitchen appliances/Gary Miller and
K.M. Scotty Mitchell.
          p.   cm.
     Includes bibliographical references and index.
     ISBN 0-87069-554-1 (pbk.)
          1. Kitchens—Equipment and supplies—Collectors and collecting—
Catalogs.   2. Household appliances, Electric—Collectors and
collecting—Catalogs. I. Mitchell, K.M. Scotty.   II. Title.
     TX656.M54 1991
     683.8′075—dc20   90-70541
                                                                      CIP
                                                                      Rev.

1 2 3 4 5 6 7 8 9 0   0 9 8 7 6 5 4 3 2 1

Contents

■

Foreword

■

Late in the summer of 1948 my parents purchased a home at 51 West Depot Street in Hellertown, Pennsylvania. Renovations to the house were behind schedule and it soon became apparent that they would not be completed by the start of the school year. My grandmother's maiden sisters, Naomi and Anna Knoble, lived at 50 East Depot Street. I was sent to stay with them so I could begin second grade on schedule. The two months I spent with them are among my fondest childhood memories.

My parents were part of the "modern" generation. Their home was equipped with a gas stove and numerous electrical appliances. As far as I knew, toast was made in a toaster and waffles were made in a waffle iron. Naomi and Anna were fiscal conservatives. In Pennsylvania German terms, they were tight with their money. Their home had electricity, but their wood/coal stove had served their parents well and still worked, so why get rid of it? The same was true of the ice box. I was only seven years old that summer, but I felt as if I had stepped back in time. Naomi and Anna made toast by sticking the bread on the end of a large pronged fork, removing one of the burner tops, and carefully holding the bread above the coals. Waffles were made in a cast iron waffle iron.

Naomi and Anna worked, so they could easily have afforded modern gas and electrical appliances. Eventually, the old wood/coal stove and ice box vanished in favor of more modern appliances, and I have always been thankful that they waited until after I left.

When talking with others of my generation, I find that my experiences were not unusual. Although electricity for household use had celebrated its fiftieth anniversary in the 1940s, a great many American homes were still using it primarily for lighting, ironing, and radio. Many kitchens still were equipped with wood/coal stoves.

For those in rural and small town America, the first decade following World War II was the "Electric Appliance Age." As though it were yesterday, I remember the arrival of our first steam iron, electric fry pan, can opener, toasted sandwich maker, and toaster oven. The electric toaster was replaced every three to four years to take advantage of the latest technology. It never left the table in the nook in the kitchen.

As an electrical appliance was updated, the obsolete appliance (which still worked) was relegated to a basement storage cabinet. One never knew when it would be needed, either in an emergency or to be passed on to an individual or couple setting up housekeeping.

When I cleaned out my parents' home in 1977, I found dozens of old electrical appliances in the basement. Alas, the collecting bug that now so dominates my life was still in the embryonic stage. Where was Miller and Mitchell's book when I really needed it?

During the past decade I have visited a number of collectors of electrical appliances. I was fascinated by the tremendous variety of design and the ingenuity of many of the working mechanisms. I quickly realized that few collectibles more adequately reflect the "average American household" and the aspirations of its occupants than electrical appliances. Needless to say, my collection now includes a number of mid-twentieth century electrical appliances.

When the antiques traditionalists attack the twentieth century collectible as devoid of craftsmanship and design, I use electrical appliances for rebuttal. Electrical appliances are the epitome of twentieth century industrial design. Their bodies changed as design styles changed. They incorporated the latest in material and technology as fast, if not faster, than the furniture industry. Some of the designs are works of art, having achieved a marvelous blend of aesthetics and utilitarianism.

Electrical appliances reflect the genius of American inventiveness. The first third of the twentieth century witnessed the growth of thousands of companies centered around a new form or refinement. Many appliances did not survive their initial marketing. Today, these failures represent some of the most sought-after electric appliance collectibles. Especially desirable are those appliances whose failure resulted from too complicated a mechanism.

Miller and Mitchell deserve credit for producing a book which will be debated and discussed by collectors and dealers. It will serve to mature the collecting category and is a most welcome point of departure.

Harry L. Rinker
Zionsville, Pennsylvania

Preface

We have tried to keep this identification and price guide simple and, we hope, entertaining. You won't find a lot of technical jargon about watts, amps, or design patents. Each chapter includes a brief history of the appliances, photos, descriptions, dates, manufacturer information, and prices. The appliances are listed alphabetically by manufacturer or, if no manufacturer is supplied, by common brand name.

We have seen drastic differences in the "asking" price of appliances across the nation. We have tried to be realistic and, in some cases, conservative in the values given in this book. It's difficult not to get overzealous when you think something is wonderful. When making a purchase, use your common sense. Because household appliances were usually made by the hundreds of thousands—or even millions—nothing in this book can be said to be "rare," though some items may be less common and harder to find.

Being inveterate collectors of the wonderful accoutrements of "kitchens past," we have literally traveled from coast to coast searching and buying, probing and questioning, and poring over countless old articles and ads (many of which are reproduced in this book).

We have enjoyed writing this price guide and hope it will be a help to collectors and dealers alike. It is amazing what you can learn by just reading the small print in many of the old ads! We hope you'll find this information as fascinating and educational as we have.

Acknowledgments

■

We wish to thank all of our friends and customers who kept on searching and finding old appliances for us. The antiques dealers in the Ft. Worth-Dallas Metroplex who have helped us in many ways also deserve a big hand. A great big thanks to our families who, instead of saying "you're doing what?", supported and encouraged us through this entire project. A special thanks to our good friends Kathryn Robinson and Vi Spencer who stood by us through it all. A big hug to you both!

We certainly must thank the staff of The Ft. Worth Public Library who rendered invaluable assistance, especially in the periodicals department, by lugging tons of old magazines to us. To Earl Lifshey, who, for the National Housewares Manufacturers Association, compiled "The Housewares Story," an invaluable thanks. It made our effort so much easier. And possibly the biggest thanks goes to Harry Rinker who verbally had to kick us in a certain place (not the kitchen) to get us to finally finish something that should have been done years ago. I don't think we would have ever succeeded without you.

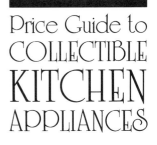

Price Guide to
COLLECTIBLE
KITCHEN
APPLIANCES

Introduction

■

For the most part, small electrical appliances are still readily available and can be found at estate and garage sales, flea markets, and auctions for little cash outlay. Even better, try raiding Grandma's attic, upper cabinets or the back of the pantry.

Generally, appliances cost very little, making them attractive to collectors on a limited budget. However, they have started turning up in antiques shops and malls, so prices are rising. This is especially true on the East and West Coasts.

Most old toasters, waffle irons, and other appliances still work. Construction is simple and basic, usually with only a two-wire connection. If the piece has an attached cord that is frayed or cracked, as in the case of the later rubber ones, it is usually no problem to remove a plate from the bottom and replace it. In the case of a detachable cord, remember that most cords came in two or three standard sizes that are still available at hardware stores today, so it is not vitally important that the cord be present unless you wish to plug it in and "try" it. Some cords are important! Nothing but an Armstrong cord, which had an unusual prong design, will work on an Armstrong appliance. There are also a few items that will require what we call a "two-headed snake." This is a single cord with two heads. If an appliance requires a special cord, don't buy it if the cord is not present. You can look for years for that particular cord.

First, visually inspect an appliance you are interested in to see if it looks intact electrically. On "flip-flop" toasters (which are the most common kind), check to see if the elements are intact around the mica and not broken. Whenever possible ask to plug in an appliance to see if it works. Do not buy an appliance that does not work, is in poor or rusted condition, or has parts missing unless you plan to strip it for parts. We had to buy four Armstrong Table Stoves before we got the waffle iron attachment. One special cord cost us about $25!

Dirt does not count. You'll be surprised to find that a little care, time and patience will clean up most appliances to a sparkling, sometimes "new," appearance. Aluminum mag wheel polish, readily available at auto parts stores and used with a soft rag, will produce wonderful results on nickel or chrome. Also, a non-abrasive kitchen cleanser can be of great help. In extreme

1

cases, oven cleaner can be used, but *be careful* (we've obliterated the manufacturer's name from appliances trying to get the old hard grease off).

In the early years of our collecting, we carefully dismantled each appliance and cleaned every part right down to the screws. This way you become very familiar with the appliance and better understand how it works. . .and you *know* it's clean. However, if you start and soon find it much more complicated than expected, *stop!* Put it back together or you may never figure it out. Complete dismantling should be reserved for those who are willing to tinker forever to reassemble the item.

As with most collectibles, the original box or instructions for an appliance can enhance the value, adding up to 25%. Original booklets impart a great deal of information and are just plain interesting.

Last, beware of chrome, silver, and other plated articles stripped to their base metal (usually copper or brass). It's really disappointing when in a shop or mall some dealer with a stripped appliance proudly proclaims "It's solid copper!" Devalue a stripped appliance by 50%.

Collecting early electric kitchen appliances can be a lot of fun. They are both usable and attractive. Because they range from primitive to shiny, high-style Art Deco designs, they make wonderful focal points and conversation pieces. Normally they don't take up much room and they require little maintenance. But, beware! If you're not careful they can multiply. Our collection is holding an entire room hostage from floor to ceiling.

In the Beginning

■

Although Benjamin Franklin has been credited with discovering electricity, we need not go back quite that far for the first electrically powered devices. In 1837, Michael Faraday was issued a patent for the first electric motor. Others contributed to the electrical beginnings in America, but there was little progress until 1882. At this time Thomas Edison established the very first electric generating plant on Pearl St. in New York City. This paved the way for electric lighting in our cities with the incandescent bulb, invented just two years earlier.

One thing that hindered the development of appliances, at least for a time, was the "battle of the currents"—AC or alternating current versus DC or direct current. This went on until about 1900 when AC won out, though it had much difficulty overcoming the short-lived DC tradition. It would have made little difference which current Edison used for lighting purposes, but he had been dedicated to DC. He was so sure DC was the way, that he personally persuaded the New York legislature to legalize the electric chair for the execution of condemned criminals, and saw to it that only Westinghouse equipment was used for the chair. George Westinghouse, the then new manufacturer of transformers and generators, was primarily interested in AC. Edison saw the electric chair as a way to dramatize the "lethal" nature of AC. AC eventually won out, and it has been said that this was the worst mistake of Edison's career. In 1908, he told George Stanley, son of the inventor of the Westinghouse transformer, "By the way, tell your father I was wrong."

Today, we have a small device that fits into a light socket and converts AC to DC. Although this somewhat reduces power to the bulb, it will last almost indefinitely. I wonder if Edison was so wrong after all?

Privately owned power companies started to spring up across the country. Electricity had been used commercially in industry for years and now it was being supplied to homes. At first, the electric power was turned on only in the evenings. It wasn't until a utility employee in California invented an electric iron that electricity was furnished to homes during the day.

Shortly after the turn of the century, things started to

1

change rapidly. Appliances of every description began to appear. One would think that the electric companies would have been the first to promote the new and innovative uses of their commodity. They were not, however; they largely restricted themselves to selling lighting fixtures.

Department stores, particularly John Wanamaker in Philadelphia, were the first to go broad scale into the business of selling the intriguing new household appliances. In October of 1906, Wanamaker staged an "Electro-Domestic Science" exposition. People were fascinated, if not somewhat skeptical of, electricity and all of the new labor saving devices.

By the 1920s, appliances were everywhere. Competition was fierce, innovations were many, and changes were rapid. This was when the ancestors of our modern appliances took shape. Through the years, the simplicity of those early inventions has changed. Today our electric food servants all but think with sensors, timers, microchips and the like. Much of the "have to" drudgery has been taken out of the kitchen—the room that we like to feel is the most important room in the house.

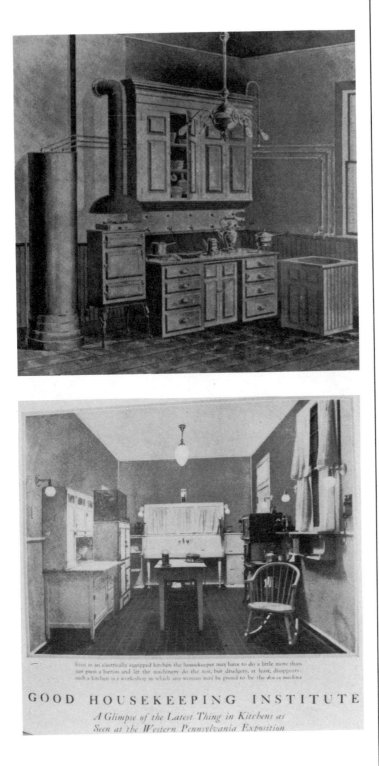

The all-electric kitchen as seen at the 1893 Chicago Columbian Exposition. This was the "kitchen of the future." It foretold of the things to come for the American housewife.

GOOD HOUSEKEEPING INSTITUTE

*A Glimpse of the Latest Thing in Kitchens as
Seen at the Western Pennsylvania Exposition*

From *Good Housekeeping*, December 1916.

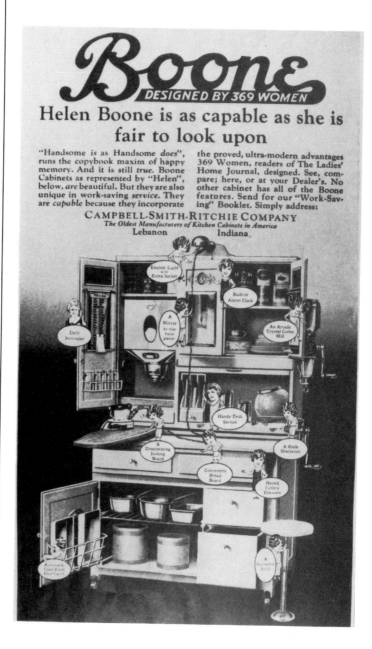

By 1927 electricity had touched everything in the kitchen including this Boone Cabinet which had an electric light with extra socket.

This *Ladies' Home Journal* article of Jan. 1936 shows toasters, cookers and warmers of the "modern" home.

This 1937 ad shows a Hotpoint all-electric kitchen featuring "Vogue" hot water heater, "Chesterfield" range, "Waverly" dishwasher and "Tudor" refrigerator.

Norge was among those manufacturers whose total production was directed toward the "war effort" in 1943. It applauded the American homemaker who could do without a new Norge until after the war and assured them that their reward was the knowledge that they were helping to speed victory and peace.

This is how a
NEW 1943 NORGE
would look in your kitchen

Startling, isn't it? But here is the new 1943 Norge Rollator Refrigerator which you are doing without.

The American behind the pair of guns can swing his turret completely around as swiftly as you can point your finger. In a flash, he can tilt his sights up to the sky or dart them toward ground or water.

No foe in air, on land, on sea is fleet enough to elude his searching aim. The target found, he can check his motion in a hair's breadth and, in the same split instant, can loose a shattering stream of fire.

Such is the new Norge for 1943. It embodies more than the actual steel and other critical materials which would have gone into your refrig-

erator. Into it have gone, too, the bold imagination, the conscientious skill, the mechanical deftness, the "know-how" which have made Norge refrigerators so fine in the past and which would have made your new Norge the finest ever built.

Your reward for doing without your new Norge is the knowledge that you, too, have helped to speed the day of Victory and Peace.

When the guns are stilled, you can be sure that Norge thinking and Norge skill, stimulated by the stern school of war, will bring you even greater satisfaction, greater convenience than you have enjoyed before.

NORGE DIVISION BORG-WARNER CORPORATION, DETROIT, MICH.

NORGE HOUSEHOLD APPLIANCES   BW

Coffee Makers
and Sets

■

The desire for the "perfect" cup of coffee is as old as coffee itself. Until the invention of the French biggin (a coffee percolator) in 1800, and for a great many years later, coffee was made simply by boiling ground coffee in water "until it smelled good." The biggin was named after its inventor and was a form of drip pot. The earliest pots were made of tin, copper or pewter. With the invention of porcelain enamel in 1875, things began to change. By the end of the century nickel-plated pots appeared.

The American people have shown the world the most development and innovation toward that perfect coffee. Electric coffee makers first appeared on the market around the turn of the century. Landers, Frary & Clark (later Universal) introduced its first electric appliance in 1908, a "Universal" percolator with a remarkable innovation called the "cold water pump." In ordinary percolators then being manufactured, the element had to heat all the water in the pot to about 190°F. before it began to perk. The Universal model was designed with a small well in the base to which the heating element was brazed, thus concentrating the heat in a small quantity of water which started to percolate in only two or three minutes. This cold water pump caught on quickly and gained fame and sales. Other models were quickly outmoded. Variations and adaptations of the pump concept soon followed. One interesting development was patented by the Metal Ware Corporation of Two Rivers, Wisconsin in 1916. This was the first immersion type heating unit. This, with some modification, is still used extensively today.

Long before any type of consumer protection groups were initiated, safety, where electricity and liquid came together, was a chief concern of many. Landers, Frary & Clark introduced in 1915 a lead fuse or safety plug which could be changed quickly and simply. Often if the appliance was left plugged in after the water became exhausted the pot would burn out. A unique gravity-operated safety device was introduced by the Robeson Rochester Company in 1924. It was described as a safety switch that not only prevented burn outs if and when the percolator ran dry, but eliminated the need to replace the blown out fuse. Be-

2

9

lieve it or not, you had to turn the coffee pot upside down and shake it to reconnect the switch.

In 1931, Knapp-Monarch introduced the first percolator designed to shut off the current after a predetermined time and temperature was reached. That same year, S.W. Farber, Inc. introduced their first coffee maker with an ingenious safety device called the "eight-in-one fuse." In 1937 (then part of Hoover), they introduced the "Coffee Robot." Featured as a coffee maker that "does about everything but buy the coffee," this vacuum-type brewer made the coffee, shut off the current when it was finished and, with a thermostat control, kept the coffee hot indefinitely.

During the 1920s and 1930s several manufacturers introduced coffee makers with separate bases which contained the heating units. The speed, simplicity, and performance of the percolator made it highly popular—so popular that it prompted many of the big utensil manufacturers to move into the appliance business and start making percolators.

Through the years design changes and innovations have been many, and sometimes a little strange to behold. Who in the early days would have dreamed that a coffee maker would have a clock to start it? In the search for that perfect pot of coffee perhaps we can look forward to a brewer that will not only turn itself on and off but feed itself water and grind and add the coffee beans as well. "Instant" coffee in my home? Never!

Manning-Bowman Coffee Percolator: Late 1920s. Manning-Bowman, Meriden, CT.

#250. Aluminum 3-part body in unique design prevents reperking. Stands 12½″ and has clear glass insert in domed lid.   **$35**

Manning-Bowman Percolator: Mid 1930s. Manning-Bowman, Meriden, CT.

Serial #636. A graceful percolator of tall Art Deco design. Stands 12″ with reeded decoration around neck and base. Part of a larger set with matching pieces.   **$40**

Porcelier Percolator: 1930s. Porcelier Mfg. Co., Greensburg, PA.

#5007. Part of a large breakfast set. Ivory porcelain body embossed with basket weave design. Multi-colored floral transfers as well as silver line decoration accent this beautiful piece. **$65**

Royal Rochester Percolator: 1920s. Robeson Rochester Corp., Rochester, NY.

Model #B29. Stands 10″. Nickel over copper body. Heavy construction. Black wooden handle. **$25**

Rome Electric Percolator: 1910s–1920s. Rome Mfg. Co., Rome, NY.

#CEU47. Coffee urn in chrome standing 14″. Wide flared black wooden handles and black turned wooden feet. Spring release on base for fuse access. **$25**

Royal Rochester Percolator: 1930s. Royal Rochester Corp., Rochester, NY.

#D30. An almost white porcelain with slight greenish luster around shoulder, handle, and spout. The porcelain is mounted on chrome base and is decorated with spring bouquet transfers on sides, handle and spout. A green line decoration further enhances the piece. **$80**

Samson-United Percolator: 1930s. Samson-United Corporation, Rochester, NY.

#155. Ivory colored porcelain body mounted on chrome base. Orange and green floral transfer decorate the piece. Most likely part of a larger set. **$60**

Silex Dripolater: Early 1930s. Hartford, CT.

Glass, 2 pc. coffee maker with glass drip insert and cupped hot plate base. Also has Bakelite holder for top. Stands 13″. **$18**

Sunbeam Coffee
Maker/Hotplate: 1930s.
Chicago Flexible Shaft Co.

Nickel-plated high Art Deco
design with black line decoration in three parts. This
with hotplate. Note revolving
heat indicator handle on
base. Ranges from 0°F to
700°F. Handles are of black
Bakelite and whole unit
stands on tab feet.   **$65**

Sunbeam "Coffee Master":
1939–1944. Sunbeam, Chi-
cago, IL.

Two-part in chrome with
Bakelite handles and base.
Stands 12½″. Controls on
base of pot.   **$25**

Farberware Coffee Maker:
1930s. S.W. Farber, Brooklyn, NY.

Model #208. Percolator in chrome with glass insert in lid. Stands $12\frac{1}{2}''$. Body has garland drape design and black wooden handle. **$20**

Farberware "Coffee Robot": 1937. S.W. Farber, Brooklyn, NY.

Coffee maker #500, Set #501. Set consists of coffee maker, creamer, sugar and tray in nickel chrome with walnut handles on pot. Two-part dripolator. "The original automatic coffee robot" advertised "no watching necessary." Also came in slightly different models. This one has original booklet.

$75 (set).

Farberware "Coffee Robot": 1937. S.W. Farber, Brooklyn, NY.

#610. Two parts with round chrome body and glass top with chrome lid and walnut handle, Bakelite knob and screw-on spout cover. Could buy as individual piece or set. Stands 13″ and would brew the coffee and keep hot automatically. Came in different sizes.

$60; **$90** (4 pc. set).

Percolator: 1920s–1930s. United Metal Goods Mfg., Inc., Brooklyn, NY.

Model #750. Stands 11½″ in pierced chrome. Design has tulip motif. Glass liner shows glowing coffee beautifully when brewing. Glass basket, black wooden handles and feet. **$45**

Tilting Automatic Coffee Maker: 1930s. "United," United Metal Goods Mfg., Inc., Brooklyn, NY.

Cat.#760. Isn't this special! The only one we've ever seen. Cast chrome decorated body and stand. Measures 16″ overall with curved, black Bakelite handle. Has red "ready" indicator light on front. Knobs on sides of pot serve as rests on pierced decorated stand. **$85**

Universal Coffee Maker: 1920s.

#E 7256. Stands 10″ in little chrome body with black wooden handle and feet. This 8 cup version came in several sizes. **$15**

Universal Coffee Maker: 1920s.

#EA 4284. Stands $8\frac{1}{2}''$ in chrome body and serves 4 cups. Black wooden handle and feet. **$12**

Universal Percolator: 1930s. Landers, Frary & Clark, New Britain, CT.

#E 6927. Part of a complete breakfast set. Body of ivory porcelain mounted on chrome base. Beautiful multi-colored floral transfer accents the design. **$65**

Westinghouse Coffee Maker: 1940s. Westinghouse, Mansfield, OH.

Model #CM-81. 14", 2 pc. maker/server in nickel with black Bakelite handles. Concentric line decoration and pierced lid. **$15**

Sets

Krome-Kraft Coffee Set: 1930s. Farber Brothers, NY, NY.

Handsome and unusual Art Deco design in nickel chrome. Tall cylindrical body accentuated by horizontal reeded area at base and has brown Bakelite handles. Set consists of percolator, creamer, lidded sugar and machine-chased tray. **$85**

Keystone Ware "Perc-O-Matic" Set: 1930s. Forbes Silver Co., Meriden, CT.

#C6602. 14″ tall in chrome with black Bakelite swing handles with red line decoration. Set has gadrooned decoration surrounding bodies. Consists of percolator, creamer & sugar.   **$60** (set).

Manning-Bowman Coffee Set: 1920s. Manning-Bowman, Meriden, CT.

Cat. #K475-9. Stands 12″. Accented by vertically mounted black wooden handles. Note the squat creamer and sugar. Set consists of percolator, creamer, sugar with lid and oval tray.   **$50**

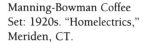

**Manning-Bowman Coffee
Set:** 1920s. "Homelectrics,"
Meriden, CT.

15″ vertically faceted urn
percolator with swing handles in ivory painted wood.
Stands on short cabriole
legs. Set consists of urn,
creamer, sugar and round
tray in nickel. **$60** (set).

**Manning-Bowman Breakfast
Set:** Late 1920s. Manning-
Bowman, Co., Meriden, CT.

Brightly finished chrome
made up of matching pieces,
not sold as a set. 4 pc. chafing dish with high/low control, waffle iron, coffee
maker and toaster (not
shown). High Art Deco styling with vertical reeding and
black Bakelite knobs and
handles. Just beautiful!

$40 (coffee maker);
$40 (waffle maker);
$45 (chafing dish);
$40 (toaster).

Meriden Homelectrics Percolator Set: Early 1920s. Manning-Bowman Co., Meriden, CT.

Catalog #32, Serial #4-30. Consists of 15" coffee urn in nickel chrome with up-turned, black wooden handles and short cabriole legs. Body has slight vertical facets. Set consists of percolator, creamer & sugar.

$60 (set).

Porcelier Breakfast Set: 1930s. Porcelier Mfg. Co., Greensburg, PA.

All-porcelain bodies accented by basket weave design in the porcelain with floral transfers. Further decorated by silver line decoration. Should be considered most unusual to have entire set.

$350. Individually:
$60 (sandwich grill #5004);
$65 (percolator #5007);
$30 (creamer, sugar and lids);
$75 (toaster #5002).

Royal Rochester Coffee Set:
1920s. Robeson Rochester
Co., Rochester, NY.

Model E 610. Lusterware
porcelain body marked
"Fraunfelter China, Ohio."
Pieces are tall, graceful and
extremely well made. Decoration consists of alternating
vertical stripes of luster and
white with floral transfers
and silver lines. Set consists
of percolator, creamer, and
sugar with lid.   **$150** (set).

Universal Coffee Set:
1912–1924. Landers, Frary
& Clark, Meriden, CT.

Classical urn shape with upturned scrolled handles in
chrome silver also has
scrolled feet. We named this
one "Brune Hilde." It has
green depression insert in lid
and spigot has "French
Ivory" (celluloid) handles.
Set consists of percolator,
oval tray, creamer & sugar.
Unique!   **$225**

Universal Set: 1910s–1920s. Landers, Frary & Clark, New Britain, CT.

#E 9219. Coffee urn stands 14″ on cabriole legs and has large black wooden handles. Four piece set consists of oval tray, coffee maker, creamer & sugar.   **$65**

Universal Coffee Set: 1920s. Landers, Frary & Clark, New Britain, CT.

#E 9119-1. Extremely graceful urn stands 16½″ in gleaming chrome. Great unity of design accented by tall handles joined to body at shoulder and just above flared base. Note tall glass insert. Set consists of percolator, creamer, sugar with lid and elongated, octagonal handled tray.   **$125**

Universal Coffee Set: 1920s. New Britain, CT.

Model #E 9189. A classic in chrome silver. Stands 14″ on three curved, reeded legs. Small decorated drop handles swing from sides. Laurel design surrounds body above legs, at shoulder and at lid edge. This is also repeated on glass insert in top. Consists of urn, creamer & sugar and rectangular handled tray with matching design. **$150**

Universal Breakfast Set: 1930s. Landers, Frary & Clark, New Britain, CT.

This handsome set consists of waffle maker, percolator, creamer, sugar and syrup. Bodies of the latter three are of cream porcelain with multi-colored floral transfers. Waffle maker is particularly handsome on nickel chrome pierced base. Has light/dark lever in front and drop handle.

$350 (4 pc. set).
Individually:
$65 (percolator #E6927);
$25 (creamer, sugar and lid);
$30 (syrup with chrome lid);
$65 (waffle iron #E6324).

Westinghouse's electric coffee maker ad of November 1917.

This Hotpoint ad in 1918 shows the Hotpoint "Grill" with "Ovenette," toaster and percolator. Even then they were trying to cook right at the table.

By 1921 Royal Rochester
percolators had a glass top
insert so you could see the
coffee perking and you
could hear the coffee "chor-
tling" right at the table.

Hotpoint was urging house-
wives to purchase a breakfast
set in this ad of March 1925.

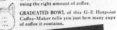

Half of a two page ad from the 1930s. This part shows the Sunbeam "Coffeemaster." Note that the late Art Deco design on the side matches that of Sunbeam toasters of the same period.

G.E. Hotpoint touted electrically brewed coffee made simply in this May 1935 ad. Note similarity to Silex.

This 1937 Silex ad shows the 2-unit, 16-cup Buffet Service. The black Silex buffet trays could be bought individually or $6.95 for a dozen.

Make Better Coffee

Want perfect coffee every time? Then—take the guess work out of coffee making. Silex glass coffee maker thinks for you ...automatically and correctly times the period of coffee brewing after switch is turned off. Pyrex brand glass, guaranteed against heat breakage. Clean...easy to keep clean.

ANYHEET CONTROL SILEX
Keeps Coffee Any Heat

Perhaps your family straggles in for coffee at different times. Now—no more cold coffee for late comers. With Silex Anyheet Control you dial the heat you prefer...keep your coffee at any heat desired...without reheating. The last cup tastes just as good as the first!

ANYHEET CONTROL SILEX **$6.95**

Other Electric Table Models **$4.95** up

Kitchen Range Models **$2.95** up

The Silex Co., Dept. 5, Hartford, Conn.
The Silex Co., Ltd., Ste. Therese, Quebec

Brewing completed without removing glass from stove

HIGH HEAT

LOW HEAT RANGE

OFF

DIAL YOUR HEAT
Anyheet Control may be purchased separately for recent electric models, $1.50

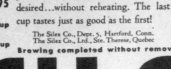

Genuine SILEX
GLASS COFFEE MAKER
TRADE MARK REGISTERED U.S. PAT. OFF.

This 1937 Silex ad shows the familiar Silex coffee maker with the new addition of a fitted hotplate and the "Anyheet Control."

Makes perfect coffee EVERY TIME
Automatically
without watching

MAN What's this?

SET IT! FORGET IT!
Shuts off by itself when coffee is done—then re-sets itself to keep coffee hot.

The Ideal Gift

Sunbeam
AUTOMATIC
COFFEEMASTER

Simply put in the water and coffee, flip the automatic switch and forget it.

Yes, that's all—*absolutely all* there is to making the most delicious coffee you ever tasted every time with Sunbeam Coffeemaster.

You don't watch it, or even think about shutting it off. It shuts itself off at exactly the *correct time every time*. You don't worry about the quality of coffee, or fret for a second about the coffee getting cold after it's made. Coffeemaster coffee is clear, mellow, full-bodied, delicious every *time* whether you make one cup or eight. No guesswork. You can't miss. Coffeemaster even re-sets itself to keep the coffee piping hot after it's made. Think of the confidence and comfort of KNOWING that every time you make coffee in your home it will be, not just ordinary coffee, but the kind that makes people stop and give compliments. No bowl breakage either—it's all gem-like chrome plate.

In justice to yourself, see Coffeemaster today. Once you witness its carefree operation, watch the marvelous coffee it makes, you'll recognize a long-awaited friend. None other like it. Coffeemaster sold alone or with matched tray, sugar and creamer set.

IT'S SIMPLE AS A.B.C.

Made and guaranteed by CHICAGO FLEXIBLE SHAFT COMPANY, 5600 Roosevelt Rd., Dept. 30, Chicago
Canada Factory: 321 Weston Rd., Inc. Toronto, Ont. Half a Century Making Quality Products

Famous for *Sunbeam* TOASTER, MIXMASTER, IRONMASTER, WAFFLEBAKER, SHAVEMASTER, etc.

Sunbeam ad of 1941 states "Set it! Forget it! Shuts off by itself when coffee is done—then re-sets itself to keep coffee hot. . . . Simply put in the water and coffee, flip the automatic switch and forget it."

"Coffee at its Best Every Time!" advertised this 1937 booklet from Farberware Electrical Appliances. Shown are four individual models of the "Coffee Robot" (this page) and six separate coffee sets (opposite page) guaranteed to "Grace Any Table."

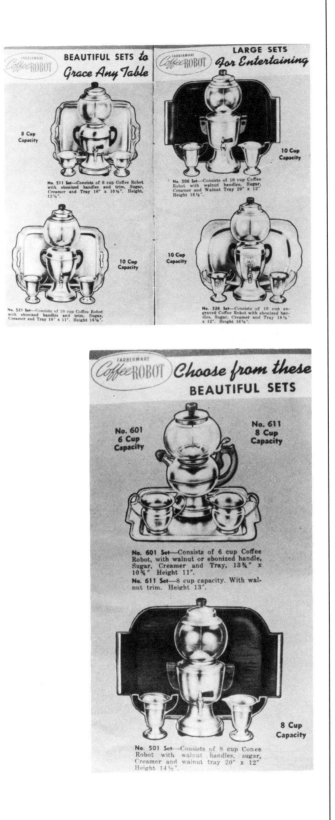

Sunbeam was proud of what it termed "America's Most Beautiful Coffee-Maker." The handsome design was three parts with hotplate base. Acted as its own server or could be purchased with service set.

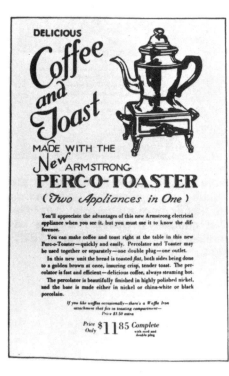

An ad for the Armstrong "Perc-O-Toaster" (two appliances in one). There was also a waffle iron attachment!

Combination Appliances·

■

Today, people spend more time in the kitchen than any room in the house. The kitchen serves as a gathering place for family and friends, and many meals are served there. Many families ate in the more formal dining room before World War II. During this era many homes had domestic help on a regular basis. Many, of course, did not. It seemed to Westinghouse, Charles C. Armstrong and others, that it would be more convenient for the American family to cook neatly and safely right at the table.

Armstrong really did something different in 1918 by introducing the "Perc-O-Toaster," a combination percolator/toaster. The coffee percolator sat atop a hotplate that radiated heat both up and down. Enclosed in a nickel cabinet on legs was a wire rack that slid out one side for the toast.

By 1916 Armstrong went a step further and introduced the "Table Stove," an ingenious device. This just about did it all! Basically it was a hotplate that radiated heat up and down in a square metal box, but it had little grooves built into the base that would hold various pans, racks, etc. There was an egg poacher (the bottom of which would double as a sauce pan), a skillet (which also had an inset broiler rack), another pan (which would make a double boiler when used with the first), a toast rack that fit into a special slot, and finally, a lid. For an extra charge one could get a waffle iron that would also fit in the toast slot. The waffle iron is very hard to find.

It wasn't long before Edison, Hotpoint, Universal and others introduced their own versions of this revolution in home cooking. They were all more or less alike, perhaps with the exception of the shape, like a round one by Edison. Some also had a high/low plug control. This heat control was very important. You could starve to death waiting for the thing to get hot enough to actually cook. Of course, it didn't spatter the good linen that was always on the table, but it was *slow*. Today these combination appliances are intriguing and a joy to own.

3

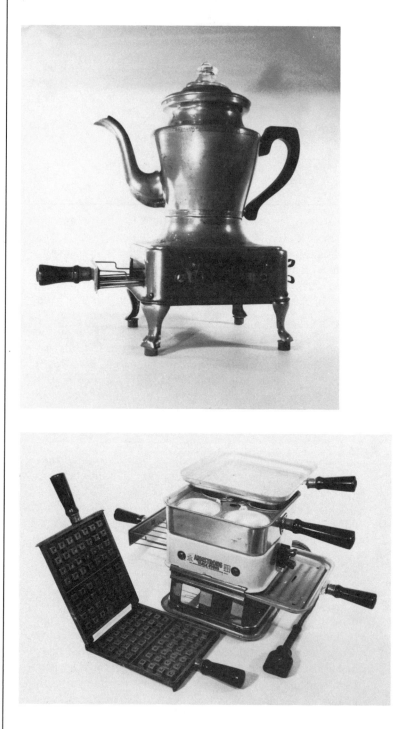

"Perc-O-Toaster"
Toaster/Percolator: 1920s.
Armstrong Electrical &
Mfg. Co., Huntington, W.
VA.

Model PT. (could it stand for
"Perc-O-Toaster?") Unusual
piece in nickel has a hotplate
base with raised center on
which fits the coffee percolator. Of important note: Base
has two special Armstrong
sets of prongs so that both
appliances could be used
singularly or in combination.
We think it's wonderful. **$60**

"Table Stove": 1917. Arm-
strong Mfg. Co., Hunting-
ton, W. VA.

The original table top stove
introduced by Armstrong.
Nickel and porcelain enamel
body with toaster rack,
broiler pan & liner (this was
also a skillet), waffle iron,
four egg poacher, and lid.
Black wooden handles and
original cord.   **$95**

"Breakfaster"
Toaster/Hotplate: 1930s.
Calkins Appliance Co.,
Niles, MI.

Model #T 2. Art Deco styling has louvered sides,
rounded corners, Bakelite
base and handles. A small
rectangular door in front
side opens to bring out attached toast tray. Top is hotplate. No controls.   **$35**

Table Top Stove: 1910s.
Edison Electric Appliance
Co., Inc., Ontario, CA. &
Chicago, IL.

Nickel body. Skillet with
handled lid could be used as
griddle, pan with four egg
poacher & broiler. Note the
pierced design on base sides
that match an Edison toaster
of the same era.   **$65**

Table Top Stove: Early
1910s. Edison Electric Appliance Co. NY, Chicago,
Ontario, CA.

$7\frac{1}{4}''$ round nickel body has
broiler with rack, pan with
lid that serves as griddle. All
rests on low base and has
four prong plug that controls
high/low heating.   **$60**

Electric Range: 1930s. Eureka Vacuum Cleaner Co., Detroit, MI.

Art Deco styling with cream painted body, black porcelainized trim, chamfered top shape, and "Eureka" across front. Bakelite handles and knobs. Interior racks. Sides fold down with round hotplates on chrome surfaces. Front panel has large red indicator light. Temp. controls for oven: Pre-warm, hi-med-low. Hotplates: hi-med.-low. Legs are part of body and top has carrying handle. This one has been restored. **$125**

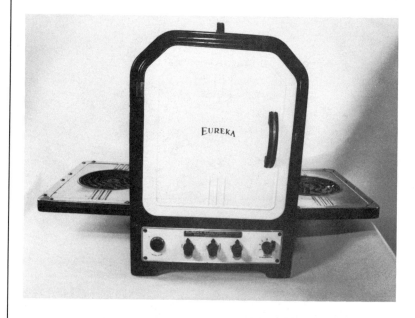

Table Top Stove: Late 1910s. Universal (Landers, Frary & Clark), New Britain, CT.

Round body measures $7\frac{1}{4}''$ and is nickel with long black wooden handles. Contains skillet, sauce pan and lid/griddle. Has high/low control and, most important, needs two headed cord. **$60**

Table Top Stove: 1922. Universal (Landers, Frary & Clark), New Britain, CT.

"New and Improved" but also is #E 988 and was actually patented two years earlier than its twin. Identical except this one has a large white china, nickel and Bakelite knob heat control. Knob is marked "Universal." **$80**

Table Top Stove: 1924. Universal (Landers, Frary & Clark), New Britain, CT.

#E 988. Shiny nickel chrome and aluminum. Black wooden handles and deep pan serves as double boiler in conjunction with bottom of four egg poacher, broiler pan and tray (skillet) and lid. High/low is controlled by four prong plug. Base has pierced design. **$75**

GOOD HOUSEKEEPING INSTITUTE

This solid-top electric stove can be used in kitchen or nursery. The luminous toaster is also practical when it serves as a grill

An upright toaster and chafing-dish are a good combination. Many dishes made in the chafing-dish are served with toast

We *Recommend* Electricity

In December of 1918, *Good Housekeeping*'s article "We *Recommend* Electricity" shows several exciting, new appliances.

"*A delightful luncheon for four and on the maid's day out. With our Armstrong Table Stove I never have to leave my place at the table.*"

Cooks Three Things At Once

It costs no more to cook all three on the Armstrong Table Stove than it does to operate the single electric utensil. The patented design of the stove concentrates all of the heat from the two heat units on the utensils so that the proper cooking temperature is quickly reached.

You can boil, fry, toast, broil or steam. A complete equipment of light, aluminum utensils comes with the stove including a griddle, deep boiling pan, toaster, four egg cups and rack.

Waffles and toast made on the Armstrong are ready in half the time for they are browned on both sides at once. No grease is necessary with the Armstrong Waffle Iron and no turning.

Ask your dealer to show you the tilting plug connection of the Armstrong Stove. The plug never sticks but lifts on or off at a touch, giving you perfect control of the heat.

Armstrong Table Stoves are for sale by most electrical supply and hardware dealers for $15.00. This includes all of the equipment mentioned above excepting the waffle iron which is $5.00 extra. Write for booklet A.

THE STANDARD STAMPING COMPANY
121-W Seventh Avenue   Huntington, West Virginia

★

ARMSTRONG
TABLE STOVE

This 1921 ad for the Armstrong Table Stove says "A delightful luncheon for four and on the maid's day out . . . I never have to leave my place at the table."

Enjoy Cool Summer Cooking

Prepare cool Summer breakfasts and luncheons right at the table with the aid of a Hotpoint Percolator and Triplex Grill. The Hotpoint Triplex Grill will broil bacon or chops, fry, boil, poach eggs and make beautifully browned toast — *any three operations at once.* You can cream chicken, chipped beef or some other light delicacy for luncheon. And the Hotpoint percolator assures you a new coffee delight. All without taking a step from the table.

See your nearest Hotpoint dealer today. Start these easy, cool Summer breakfasts and luncheons tomorrow.

(The Triplex Grill is $13.50; others from $4.00 up. Percolators from $9.00 up.)

EDISON ELECTRIC APPLIANCE CO., Inc.

World's largest manufacturer of electric ranges and household electric heating appliances

5600 West Taylor Street, Chicago

Factories: Chicago, Ill., and Ontario, Calif.
Branches and Factory Service Stations
in principal cities

(In Canada: Canadian General Electric Co.,
Ltd., Toronto.)

★

SERVANTS

The Hotpoint "Triplex Grill" advertised in 1927 that you could perform three cooking operations simultaneously right at the table.

Star-Rite's Christmas ad in 1927 shows a variety of their appliances. Note their home motor that sold for $20 at the time; nearly four times the amount of the toaster and double the price of the waffle iron.

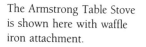

The Armstrong Table Stove is shown here with waffle iron attachment.

Thousands of happy homes have been enjoying the Armstrong Table Stove for years. It's something every woman loves to use— and it's such a help at breakfast, luncheons, and for dainty servings at any time. Cooks three things at one time—makes waffles too. Always an appreciated gift. For sale at good electrical and hardware stores.

Cooking Appliances:
Broilers to Poppers

■

This chapter will center on those early appliances that were designed solely for cooking food. These range from hotplates and slow cookers to egg cookers and popcorn poppers. Hotplates have been with us nearly as long as the iron. The early ones have the elements encased in a clay-like compound below a steel plate and raised up on little legs. We don't usually think of the iron as a cooker, but we're probably not the only ones who used an iron for cooking while in college. We weren't allowed to have hotplates in the dorm, so the only way around this was to use an iron upside down between two bricks. It worked just fine . . . until we got caught.

Broiling food is thought to be the oldest method of cooking (in one form or another). Electric broilers made for the table began to appear in the U.S. early in this century and were very popular. Previously, fried or roasted foods were preferred. As has always been the case, nutritional values and the type of food we eat change with the times. Rightly or wrongly, fried foods were blamed for many ills and fell into disfavor after the late 1920s or early 1930s. Sixty years later, broiling remains a popular form of cooking.

The first broilers for table or countertop use consisted of a base on feet with a wire rack and some sort of domed top which housed the heating element. It wasn't until about 1937 that Manning-Bowman came out with a broiler that "did not smoke like the others." Although it was designed with a high, round domed top that fit snugly on the base and had three tiny holes, it had the same problem as many others—poor air circulation. In 1938 the Puritan Aluminum Company, which would later become the Dazey Products Co., marketed an "electric steak platter" which was cast aluminum with a well-and-tree recess and an electric heating element underneath.

Broilers have certainly improved over the years from the primitive turn-of-the-century broilers to the browning elements in our indispensable microwave ovens.

Popcorn has been around since the Pilgrims enjoyed it at the first Thanksgiving. The first poppers were probably the "wire over the fire" devices made by the Bromwell Products division of Leigh Products in 1819. The first electric corn popper

4

appeared in 1907. A magazine article of that year states: "Of the host of electrical household utensils, the new corn popper is the daintiest of them all. Attach the connection with the electric light socket, and the children can pop corn on the parlor table all day without the slightest danger or harm ... The results are far better than the old way of building a red hot fire in the kitchen range and suffering from the heat while popping corn." The article does not mention a manufacturer.

The early poppers were very simple affairs that were basically enclosed hotplates in which fit a container with a hand crank on top. Our earliest is unmarked and has a wire basket on top of a tin can. How primitive can you get? It's electric but that's about all! We think it may well be the first.

Mirro Vice-President Frank Prescott once remarked that popcorn would become a breakfast favorite. Popcorn is popular, but so are eggs. Early in this century Hotpoint had "El Eggo" along with "El Perco," "El Grillo," "El Tosto," "El Chafo"—well you get the idea. Of "El Eggo," an ad states: "An electric egg cooker. Boils, poaches & scrambles eggs. May be used right on the breakfast table. A most unique and useful device. Price $9. In Canada, $11.95." An interesting and pretty thing of bright shiny chrome, it looks like a big egg on a base with black wooden handle and knob. It has the little egg holes inside and even a skillet and looks identical to one made by Rochester.

There were many egg cooker designs, but perhaps the most clever innovation was introduced by the Hankscraft Company of Madison, Wisconsin (now a part of Gerber). First patented in 1921, the Hankscraft "Egg Cooker" was unique. The china base contained an enclosed heating element. Into this base was placed from $1\frac{1}{2}$ to 5 teaspoons of tap water. When the water was boiled away, the unit shut itself off and the eggs were done to the desired degree based on the amount of water. Instructions for cooking and cleaning were always present on a metal plate under the base. This also appeared on some models as a metal tag attached to the cord.

Even slow cookers were around in the early days of electrical appliances. The "Everhot" cooker appeared in 1925. This was made by the Swartz Baugh Mfg. Co. and was a cylindrical affair of bright chrome and black painted metal with insulated aluminum lid and wire handle that locked down. The interior was fitted with a rack, two semicircular, open containers and one round, lidded container. There was even a handled rod for lifting out the hot containers. A three prong plug in the base enabled the user to choose either high or low heat, depending on whether the middle and right prongs or the middle and left prongs were used. There were other manufacturers of these early slow cookers and the designs varied somewhat.

From the first crude beginnings, cooking appliances of all types have remained popular over the years. Today, some even weigh the food, calculate cooking time, temperature, and even start and stop themselves. What will they be doing in another hundred years?

Farberware Broiler: 1920s.
Brooklyn, NY.

Shiny chrome domed design
has element in top. Base has
fitted pan, rack and little
round feet. Each piece has
black wooden handle. Three
prongs with "special" plug.

$20

Electric Broiler: 1930s.
Royal Master Appliance
Co., Miriam, OH.

Model 1-A. Heavy cast
"hammered" aluminum has
base that is a well-and-tree-
platter with black Bakelite
handles. Pan and rack. Heating element in top will run
1400 watts! Large oval shape
20″. Top will stay open easily. Has three prongs set tri-
angularly. Will run on AC or
DC. Early for this type. **$25**

Chafing Dishes

"American Beauty" Chafing Dish: 1910. American Electrical Heater Co., Detroit, MI.

Three parts of nickel. Sealed heating element in the base serves as hot water container. Also has two separate plugs marked "fast" and "slow". Knobs and handles are black painted wood. **$50**

Electric Chafing Dish: 1930s. Manning-Bowman, Meriden, CT.

Elegant Art Deco design in bright chrome. Several other Manning-Bowman appliances matched this design. In four pieces, this has heating base with high/low plug designation, hot water pot, interior cooking pan and lid. Parts have black Bakelite handles.
$45

Chafing Dish: 1910s. Universal (Landers, Frary & Clark).

Nickel on copper. Faceted body design in three parts with sealed element in base and hot water pan in which the food cooking pan fits neatly inside. Three prong heat adjustment in base and black wooden handles and knobs. **$50**

Egg Cooker: 1920s. Hanks-craft Egg Cooker.

Model #599. Yellow china base and large dish on top of chrome lid that serves as a knob and also as a filler, with hole in bottom. Not only has instructions on underside but also a metal tag giving same attached to cord.   **$25** (complete).

Egg Cooker Set: 1930s. Hankscraft Co., Madison, WI.

Model #730. Beautiful Art Deco design in ivory china with silver decoration. Lid has finial knob. The set has cooker, four egg cups and oval chrome tray. Elegant at the breakfast table.   **$50**

Egg Cooker: Late 1930s. Hankscraft Co., Madison, WI.

5¼″ in diameter, this little gem has cream china base and aluminum top with matching china knob. Little Art Deco, reeded design on base and knob. Three parts; lid, egg holder and base.

$15

The underside of a Hankscraft Egg Cooker. This shows the metal plate with cooking and cleaning instructions. "You time the eggs by amount of water. A little practice tells just how much for your taste." "Suggested teaspoons of water. Soft Boiled, 2; Medium Boiled, 4; Hard Boiled, 6." These have been around since 1915.

"Rochester" Egg Cooker: 1910s. Rochester Stamping Co., Rochester, NY.

This handsome piece in chrome is egg-shaped and in four parts. The interior is fitted with skillet that has nicely turned black wooden handle. Into this fits the holder for six eggs with lift out handle. Heating element is enclosed. This is almost identical to Hotpoint's "El Eggo."

$40

"Nesco" Electric Casserole: Early 1930s. National Enam. & Stamping Co., Inc., Milwaukee, WI.

9″ diameter. Original fore-runner to the crockpot. (This one is not in the best of condition.) The fruit de-cals are on the other side. Body is cream colored and the piece has green enamel lid. High/low control and three prong plug.

$25 (in better condition).

"Quality Brand" Food Cooker: 1920s. Great Northern Mfg. Co., Chicago, IL.

Model #950. Stands 14″ with cylindrical body. Insulated sides and lid help to contain heat from the 500 watt element. Interior is fitted with two lidded aluminum pots. A lift out rod helps protect fingers. Small screws on body top secure the lid. Handle is of heavy wire with black wood. Body is brown with red stripe. Plug but no heat control. This could create some interesting comments at your next covered dish social!

$40

Hankscraft Food Cooker: 1920s. Hankscraft, Madison, WI.

Works much like the egg cookers in that cooking time is based on the amount of water put in base. Turns itself off when water is gone. Green enamel bottom pan sits on chrome base. Interior pan holds food above the water much like a double boiler. Chrome top has pin hinged lid that can be completely removed and knob is green lusterware. Black wooden handles flare from the sides. Magnificent engineering design.   **$50**

"Everhot" Food Cooker: 1920s. Swartz Baugh Mfg. Co., Toledo, OH.

The EC "Junior" 10. What size must the "Senior" be? This stands 13″ and has chrome and black painted body. "Everhot" emblazoned across the body in embossed letters. This is the original slow cooker and is probably the best known brand. Be prepared to lift this one. The interior is fitted with rack; two open semi-circular pans; one lidded pan and has a lift out handle. Three prong plug controls high or low heat.   **$50**

Edison/Hotpoint Hotplate:
1910s. New York, Chicago
& Ontario, CA.

This was the one that is responsible for our collection.
Very heavy and clay filled.
Has china knob on copper
heat control. Little pierced
legs rest on china feet. **$25**

"El Stovo" Hotplate: 1910s.
Pacific Electric Heating Co.

Can also be found marked
"G.E./Hotpoint." This is similar in design to the Edison
variety but has no control.
Flat prongs and part of the
"El" group of appliances; "El
Perco," "El Eggo," etc. This
one is early.   **$15**

**General Electric "Disk
Stove":** 1920s.

Measures 6″ in diameter and
has clay filled element chamber, china knob and
high/med./low control on
front. Notice the close similarity to the Edison/Hotpoint
version.   **$20**

Handyhot "Victory" Hotplate: 1920s. Handyhot, Chicago.

Black and very basic. Looks like it was left out in the sun until it was a little overdone.
$5

Universal "Thermax" Hotplate: 1920s. Landers, Frary & Clark.

Inner and outer surfaces of iron top barely connected by a swirl design. Held up by tripod base and has 4 prongs to heat only outer ring, inner circle or both. Wooden handle. A tiny plaque below prongs directs how to connect for desired heat. Requires special cord with two heads. This is a great piece.
$25

Universal Hotplate: 1920s. Landers, Frary & Clark.

This little well-built unit has three prongs with "fast/slow" guide. Reeded body and decorated legs. Two wooden handles make it easy to move while hot.
$20

Unmarked Hotplate: 1920s.

Model in nickel chrome measures 9″ square with no control. Stands on slightly angled legs. Very basic.   **$8**

Unmarked Hotplate: 1920s.

Round. Stands on slight ca-briole legs. Nice but plain design. Porcelain insulator where wire leaves body. **$10**

Unmarked Hotplate: 1930s.

Light weight, stamped nickel body $6\frac{1}{2}″ \times 5\frac{1}{4}″$ with rounded corners, open grid top and rounded legs. No controls.   **$8**

"Volcano" Hotplate: 1930s. Hilco Eng. Co., Chicago, IL.

This is unusual. Slightly conical body with black wooden handle contains heating element. Although no controls, a side lever mounted in notches on side lifts grate for hot or not so hot. Our lever is missing.

$20 (in good shape).

Westinghouse Hotplate: 1920s. Westinghouse, Mansfield Works, Mansfield, OH.

This pretty little $7\frac{1}{2}''$ round hotplate has green porcelain metal top surrounding element. Base is hexagonal and stands on rounded, hollow legs. No control. **$25**

Popcorn Poppers

Popcorn Popper: 1930s. Berstead Mfg. Co.

Model #302. Square chrome body encloses the hotplate with the interior being circular. Of important note is the Fry glass lid that adorns the top. A large black knob on top is attached to rod for stirring. The chrome body on our's is a little rough.

$45 (in good condition).

Challange Popcorn Popper: Early 1920s. Challange Mfg. Co.

Black bodied popper has very short legs, turned black painted wood side handles and stirrer knob. No lock down on this early one. I'll bet it made a mess. **$20**

Popcorn Popper: 1920s. Dominion Elect. Mfg. Co., Minneapolis, MN.

Style #75. This one is interesting. One piece cylindrical, nickel, pierced body with lock-down lid. It stands on little cabriole legs. Turned wooden handles and knob atop hand crank are painted red. **$25**

Popcorn Popper: 1920s. Excel Electric Co., Muncie, IN.

One piece nickel body. Metal handles form two of the legs. Handles also have little lock-down levers to keep lid from blowing off. Hand crank has black wooden knob. Top with vent holes has embossed in large letters: "Excel Electric Corn Popper." **$25**

Popcorn Popper: 1920s. Knapp Monarch, Belleville, IL.

Steel body in one part with lock-down lid and large black, wooden knobs and stirrer knob on top. **$15**

Popcorn Popper: Late 1930s. Knapp Monarch, St. Louis, MO.

All aluminum in two parts with domed glass lid (the glass cabinet knob is hardly original). Stands on heavy wire base. This one came with aluminum measuring cup. Has large, turned walnut handles. **$15**

Popcorn Popper: Late 1930s–Early 1940s. Manning-Bowman Co., Meriden, CT.

Model #500. Detachable, large aluminum container sits atop chrome hotplate. A glass lid decorated with embossed floral motif also sports black Bakelite knob. This one has never been used. **$15**

"White Cross" Popcorn Popper: Late 1910s. National Stamping & Electrical Co., Chicago, IL.

One of the very earliest corn poppers. Consists of a tin can with a wire just sticking out of the side and connected to the hotplate inside. A wire basket fits into the base and has black wooden handle. Top stirrer is mounted through a handle. Top does not lock down. This is about as primitive as you can get and still be electric. We love it. **$30**

Popcorn Popper: 1920s. Rapaport Bros., Inc., Chicago, IL.

This interesting piece has a Bakelite base measuring $5\frac{1}{2}''$ square and stands on metal legs. Upper part is aluminum with attached lid and red knob. Handle of chrome "squeezes" through a slot in the side to agitate corn. Most unusual. **$25**

Popcorn Popper: 1930s.
U.S. Manufacturing Corp.,
Decatur, IL.

Like most of the poppers of
the era this varied little in
design. Hotplate base has
detachable cord and hand
crank. Tan with orange
knob, handle, and "loop"
legs. **$20**

Popcorn Popper: 1930s.
U.S. Manufacturing Corp.,
Decatur, IL.

This is #1; the first popper
made by the U.S. Mfg. Co. It
is all in one unit and is red
and silver with black knob
holding the crank. Legs are
of wood. **$20**

Popcorn Popper: 1930s.
U.S. Mfg. Corp., Decatur,
IL.

Model #10. Probably the
most common we've found.
Three parts with lid and
hand crank. Came in several
color combinations including
tan & brown, cream &
green, and red & ivory.
These could match the color
scheme of your modern
1930s kitchen. **$15**

This "Simplex Table Range" was featured in an article in the March 1914 *Good Housekeeping*. At that time it cost 6¢ to run and it took twenty minutes to heat the three-cup capacity stew pan full of water to 200°F.

$60 (complete, current value).

This 1917 Manning-Bowman ad stated that MB appliances were designed to be used either with electricity, alcohol or on an ordinary range. Note the new electric chafing dish that year.

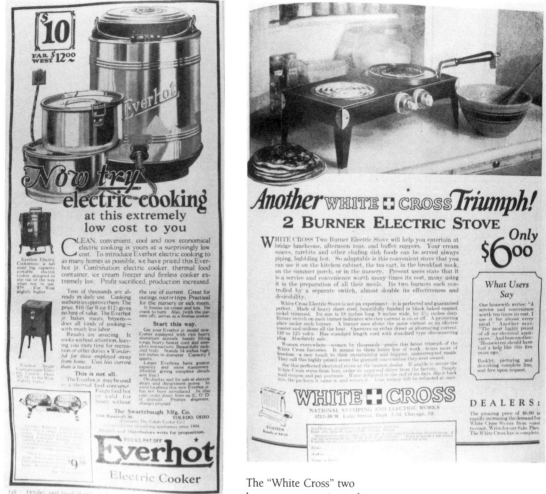

The "Everhot jr." electric cooker/ice cream freezer sold for $12 in 1925. It would cook while not attended and claimed that it could "freeze ice cream and ices without a crank."

The "White Cross" two burner stove as pictured in an October 1925 ad which claims "Another White Cross Triumph!" "Women everywhere—women by thousands—praise this latest triumph of the White Cross factories."

White Cross introduced their new all-nickel body double stove in this 1927 ad while still offering the all-steel model of years earlier.

In 1937, General Electric advertised this "Roaster Oven" and boasted that it would not heat up the kitchen and that you could cook a whole meal at one time.

$35 (roaster); **$60** (complete).

Just plug it in

AND COOK TO PERFECTION!

THE UNIVERSAL ELECTRIC OVEN

The November 1937 *Better Homes & Gardens* contained this ad for a Universal Electric Oven.
$75 (current value).

BROILRITE

THE ORIGINAL ELECTRIC BROILER PATENTED 1934

DON'T BUY INFRINGEMENTS

KIMMEL SALES CORP. ROCHESTER, N.Y.

According to this ad, the "Broilrite" was the original electric broiler. **$25**

Irons

■

The tedious chore of trying to de-wrinkle woven fabrics goes back to at least the tenth century when Vikings used a device resembling a flattened, inverted glass mushroom to smooth linens. The Greeks used a "goffering" iron to pleat their robes. The first improvement was the use of heat, but it's unknown when or where this occurred. By the sixteenth century, ironing in Europe was done with a hot charcoal-filled, hollow box-like affair. The charcoal was eventually replaced by ingots of heated iron. Later, as the fireplace was replaced by cookstoves, the sadiron came into being. After the machine age was under way irons were powered by alcohol, gasoline and even acetylene.

It wasn't until June of 1882 that a patent for the first electric iron was granted to Henry Seely of New York. This patent was not much of an improvement over previous designs and the iron was of almost no practical use. The current was only supplied while the iron sat on its stand and there were really no power companies to supply electricity until the 1890s. (There were other patents for early electric irons, which were among the first electric appliances to change the American way of life.) When the power plants did come, they were interested only in supplying electricity for generators and the fascinating (but not trusted) electric lights. These early power plants were individually owned and operated, supplying electricity only from dusk to about ten P.M.

It never occurred to anyone that power companies could prosper by encouraging the use of electric appliances until 1903. Earl Richardson, plant superintendent of the power company in Ontario, California, had been giving serious thought to this. He had vested interests for his concern, having been experimenting for some time with an electric iron. Also, if he could convince women to use irons in the daytime, then perhaps he could also persuade the power company to operate around the clock. He refined his model by making it lighter and smaller. He made several dozen samples and distributed them to customers on his meter reading route. These irons were a great success with the ladies. Also, the company agreed to supply power on Tuesdays so the customers could use the irons. The demand grew so much that the following year Rich-

5

ardson, with four employees, left the company and without financial backing formed the Pacific Electric Heating Company.

Trouble was just ahead. The demand dwindled. When he questioned the ladies using his iron, they all agreed that his was better than the rest but it had one major fault; it overheated in the center. While talking to his wife about the problem, she suggested that he redesign the iron with more heat in the tip, where it was needed to iron around ruffles, pleats and button-holes. This he did. Again, he made up samples and replaced the irons with the new model. When he went back to check, no one wanted to give up the new iron with the hot point. That was it! Not only did he have a successful product, but also a name! In 1906 Hotpoint produced more electric irons than any other company in America.

As technology and demand grew, many innovations emerged. Perhaps one of the most important of these was the introduction of the first "automatic" iron in 1924. In 1927 the Liberty Gauge Instrument Co. (later Proctor Silex) introduced an iron with adjustable heat settings. The first steam iron was made in 1937 by the Steem-Electric Co. and what has been called the steam iron "holy war" ensued. One hole in the iron just didn't seem to effectively annihilate wrinkles. Soon there were three holes. One manufacturer would double the amount of holes in a competitor's iron, and another would double that amount until finally in 1971 Presto introduced an iron with eighty holes.

Other innovations included the cordless iron from Nocord Co., 1922; an iron that could work backward without catching on pleats from Landers, Frary & Clark, 1923; the "Button Nook," 1929; and an iron that weighed less than six pounds from Sunbeam, 1930.

Irons have come a long way. Women no longer need the biceps of a blacksmith nor the endurance of a crazed marathon ironer, but they still have to hold on. Maybe some day an iron will be invented that moves by itself.

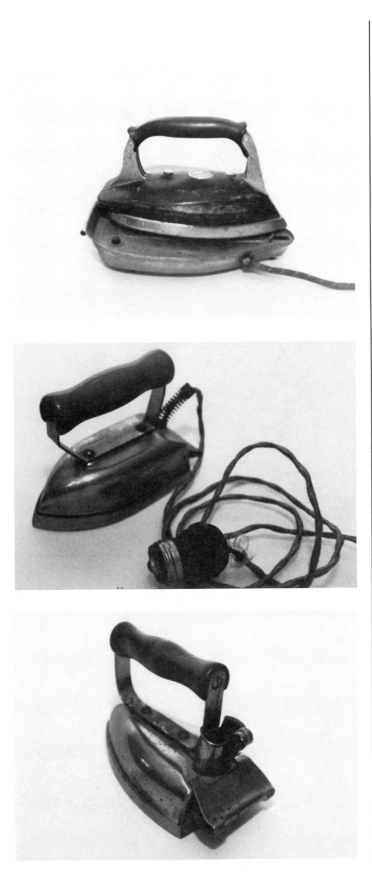

"Cord-Less-Matic" Iron: Late 1920s. Brannon Inc., Detroit, MI. Pat. Pend.

Black body with triple plates running upward into Bakelite handle. Heat indicator, no adjustment. Two points on sole plate connect with heating base element. Odd silhouette.

$25 (in good shape).

Dover Iron: 1930s. "Co-Ed," Dover Mfg. Co., Dover, OH.

$2\frac{1}{2}$ lb. with screw-in permanent cord and green wooden handle. Sometimes referred to as a child's iron. **$20**

Edison Iron: 1906. Edison Electric Appliance Co., Ontario, CA.

Very plain with nickel body and wooden handle, detachable cord. Nothing fancy here. **$5**

General Electric Iron: Late 1930s.

Chrome body with parallel line decoration. Heat indicator and button nooks, black Bakelite handle with thumb rest. A real comfortable iron.

$12

General Electric/Hotpoint Iron: Early 1930s. "Calrod-A-C-Matic."

Nickel body with red wooden handle and Bakelite thumb rest. Detachable cord, button nooks, heavy.

$10; $20 (with box).

General Mills/Betty Crocker Iron: 1940s. "Tru-Heat Iron."

Automatic 3¾ lb. with 7 holes, interesting side rest, steam attachment, red Bakelite knobs in black handle. Original box. Detaches from steam base for regular use.

$30

Heatmaster Iron: 1920s.

Cat. #372. 6″ nickel body. 2½ lb. with wooden handle. Operates on AC or DC. No other information is available as to manufacturer.   **$5**

Hotpoint Iron: 1920s. "Calrod Super Automatic." Edison, General Electric Appliance Co., Inc., Chicago, IL & Ontario, CA.

Heavy nickel shell over iron with simple control and black wooden handle. Original box just barely there. Detachable cord.   **$12**

Knapp Monarch Steam Iron: Early 1940s. "Steem-R-Dri." St. Louis, MO.

10 hole, heavy chrome, fat body with pointed back. Off/on steam switch. Bakelite filler cap and handle. Attached cord. Interesting design resembles a tug boat.
      $18

**Knapp Monarch Iron:
1930s. St. Louis, MO.**

Fish-shaped nickel body
with curved brown Bakelite
handle. Round heat control.
Impressive bronze plate on
rear of handle. Permanent
cord.   **$15**

**Knapp Monarch Travel
Iron: Late 1930s. "Gad-A-
Bout."**

Chrome, flat body with fold-
ing cream and brown Bake-
lite handle, heat indicator,
detachable cord and brown
zippered cloth bag. Never
used.   **$12**

**Kwikway Iron: 1920s.
Kwikway Co., St. Louis,
MO.**

2½ lb. nickel clad iron with
heel rest and black wooden
handle. Very plain with detachable cord.   **$8**

Petipoint Iron: Late 1930s. Waverly Tool Co., Sandusky, OH.

Model W 410. High Art Deco styling characterized by horizontal layered fins. Back sole plate tilts up. Red control lever & black Bakelite handle. Looks like it should fly! **$35**

Steam-O-Matic Iron: 1931–44. Waverly Products, Inc. Sandusky, OH & Bridgeport, CT.

Hammered aluminum Art Deco designed body with temperature control and handle in black Bakelite. Original box, Bakelite funnel and booklet. Barely used. Reminds me of an old Hudson. **$20** (complete).

Sunbeam Iron: 1940s. "Ironmaster."

$2\frac{3}{4}$ lb. iron with steam attachment. Red "wheel" control on front of black Bakelite handle. Never used, original box with papers. **$20** (**$60** with attachment).

Sunbeam Iron: 1920s. Chicago Flexible Shaft Co., Chicago, IL.

This plain little iron with its personalized hot plate has always been a favorite partly because of the green metal box that has front and top hinged opening, cord compartment and decal proclaiming proudly "34 years of quality." **$40** (complete).

Universal Iron: 1910s. "Wrinkle Proof."

Just plain-Jane with nickel body, black wooden handle, and detachable cord. That's all folks. **$5**

Westinghouse Iron: 1920s. "Adjust-O-Matic."

Nickel body with heat control, will "sit up." Has black wooden handle and detachable cord. **$8**

This November 1916 *Good Housekeeping* ad proclaimed the Westinghouse Iron would cut out "the changing of irons" and "tiresome trips to and from the stove."

This Sunbeam iron (1925)
which came with hotplate
and "all steel, fireproof box"
could be put away right after
ironing.

Chicago Flexible Shaft (later
Sunbeam) thought simple
was better when they advertised "The Domestic" iron in
December of 1920. ". . . no
frills, furbelows or dewdads."

The "Wrinkleproof" iron that had no sharp corners and ironed easily forward, backward or sideways was featured in this 1925 Universal ad.

This *Ladies' Home Journal* ad of 1925 shows my favorite Sunbeam Iron. I think it's nifty, just because of the green metal box.

This Hotpoint ad of 1929 featured an iron with a "throttle" you could set for any heat, which would remain constant. The "Calrod" heating element was "indestructible" because it was cast in solid iron, a Hotpoint "exclusive" feature.

Before the crash of '29 almost everything was becoming "automatic," including this Sunbeam iron of that year. It featured overall heating and an air cooled handle. At that time it was called the "Master Automatic Iron," the percursor to the "Ironmaster" and other "Master" appliances.

This ad from 1935 for the Proctor "Snap Stand" iron proclaimed that the dealer would buy back your old iron for $1.

In 1941 Proctor introduced the "Never Lift" iron which had a spring loaded, built-in pop-up stand that was activated by a button. It lowered at a touch. Innovative, and it worked! We have one.

...but you'll be back again!

This "Ironrite" ad of 1943 illustrates how many appliance manufacturers switched from domestic application to making gun parts during the war.

When Sunbeam cut the weight from 6½ lbs. to a 3 lb. "automatic" iron, no one dreamed we could iron with the "feather weights" of today.

Mixers and Whippers

■

Although the complete details may never be known, it is a fact that both the electric mixer and the creation of the malted milk originated in Racine, Wisconsin, around the turn of the century.

Through a purely circumstantial chain of events came the revolutionary "universal" home motor. The Arnold Electric Company, an early manufacturer of appliances in Racine, owned by Fred Osius (Osterizer) and George Schmidt, hired a farm boy named Chester A. Beach and a former cashier named L.H. Hamilton. Hamilton and Beach proved to be very important to the appliance industry. These two men developed and perfected a high-speed, lightweight "universal" home motor.

It should be pointed out that the generating plants at that time produced either AC or DC, and this motor would operate on either. Before this, a small, high-speed motor was known only in theory. This one ran at 7,200 RPM and eventually at 10,000 RPM. It has been related by the American Housewares Manufacturers Association that, "This was not long after Westinghouse had told a new manufacturer of vacuum cleaners named Hoover that it would be unsafe to attempt a motor with a speed greater than 1,700 RPM."

Hamilton and Beach were not satisfied with just a drink mixer for their motor so they developed a series of uses for the "universal" motor, from running a treadle sewing machine to sharpening knives, buffing silver and even mixing cake batter. Although I would not dream of using such an ominous device, we have even seen design patents for an application of the motor to brush your teeth! This operated on a long flexible shaft. I just don't think I could trust it!

Hamilton Beach was the first company to market a mixer mounted on its own stand in 1920. This was the KitchenAid model. The Air-O-Mix "Whip-All" was the first combination base-mounted and portable mixer, introduced in 1923. It was a vertical affair with handle attached to the motor over a long beater, all of which rested in a heavy wire stand.

In 1927 the "Dormeyer" electric "Household Beater" (which was then manufactured by the MacLeod Manufacturing Company of Chicago) was designed so that the motor could be

6

readily detached from the bracket holding the beater blades. Simply detaching the beaters did not come until later.

Some of the early mixers were sold in large quantities as premiums. In 1929, Gilbert (maker of the famous but long gone "Erector Sets") produced the "Polar Cub" hand mixer that could be used apart from its stand especially for the Wesson Oil Snowdrift company. This had a list price of $11.95.

Probably no company has contributed more to mixers than the Chicago Flexible Shaft Company (now the Sunbeam Corporation). Sunbeam introduced an entire line of "Master" appliances; the "CoffeeMaster," the "IronMaster," the "ShaveMaster," and last but not least, the "MixMaster." (They couldn't use "ToastMaster"—that name had already been taken.) The MixMaster was not only mounted on a heavy cast base but came with a juicer attachment and two stainless steel mixing bowls for which there was a ball-bearing turntable. This was offered for under $20, considerably less than Hobart's KitchenAid.

"Mixmaster" has become an almost generic term for mixers in general and Sunbeam took attachments to the limit. By 1936, at the height of a scramble to accomplish nearly every domestic task in the kitchen, the "Mixmaster" had attachments to grind and chop, slice, shred, grate and crush ice, juice and make mayonnaise, peel potatoes, mix drinks, grind coffee, open cans, sharpen knives, buff silverware, turn the ice cream freezer, rice and puree, and even shell peas. It seemed logical that there had to be a cabinet to hold the mixer and all of its attachments. It stood $60\frac{1}{2}''$ high and was 24″ wide. There was a 20 × 24″ porcelain draw-out table top. For convenience, the cabinet was fitted with an electrical outlet on the left and the whole affair plugged into a wall outlet. Not quite as compact as our food processors of today.

If asked who invented the blender most people would probably say "Waring" or "Fred Waring," famous leader of "The Pennsylvanians." Actually the prize goes to Stephen J. Poplawski of Racine, Wisconsin. From 1915 and for the next half-century, Poplawski was engaged in the design and manufacture of beverage mixers. In 1922 he applied for a patent which had an agitating element and driving motor mounted in a base with a recess to hold a cup on top. He was awarded the patent but was not thinking of using the machine for liquifying fruits and vegetables. He was after the commercial drink market. Remember this was in Racine, Wisconsin, home of the malted.

Fred Osius moved to Florida and then back North during the Depression. He had an idea for a blender, but was rather unorthodox in his ideas of investors. One afternoon in 1936 he pushed himself off on Fred Waring, who had just finished a broadcast for Ford Radio. He had brought a prototype and explained how it would revolutionize people's eating habits. He got Waring's attention. Six months and $25,000 later, Osius still hadn't produced the promised blender. In September of 1936 Waring asked an associate to take it over. Some basic engineering problems were solved in time to introduce the "Miracle Mixer" the following fall at the National Restaurant Show in Chicago. It was featured as a new method for making frozen daiquiris and was enthusiastically embraced, so much so that Ron Rico Rum had a national promotion to popularize frozen drinks. By now it was being called the "Waring Blendor" (spelled with an 'o' to distinguish it). Consumers were educated to the fact that the new "blendor" was equally good in the kitchen as well as a bar.

World War II brought the manufacturing of blenders and other appliances to a halt. All activities were directed to the war effort. Production was started again after the war. Many designs for appliances did not change. Therefore it is sometimes difficult to correctly date an appliance. Those from the 1930s and 1940s look just like those made well into the 1950s. This is particularly true of the small, high speed whippers; those cute but sometimes less than adequate devices. Many just made little whirlpools in the bottom of the glass cup. A number of companies both great and small produced the little whippers—most have now gone by the wayside. Their efforts, and those of other kitchen appliance manufacturers, have remained to enchant collectors and individuals who enjoy the simplicity and quality of designs of a bygone era.

Mixers

Gilbert Mixer, "Polar Cub":
1929. A.C. Gilbert Co.,
New Haven, CT.

Stands 10″ in gray painted
metal, has rear switch and
bright blue handle. The
mixer lifts off holder. Sold
for $11.95 as a premium ex-
pressly for Wesson Oil
Snowdrift. Smaller models
with glass bowls sold for
$4.95 and $7.50. This one
still has the original box.
  $75; **$60** (without box).

Gilbert Malt Machine:
1930s. A.C. Gilbert Co.,
New Haven, CT.

13½″ model operates on AC
or DC with chrome motor
and rear off/on switch. Base
is a green painted, crinkle
finish over cast iron. We've
not yet found the cup for
this one.
  $65; **$75** (with cup).

Berstead Drink Mixer: 1930s. "Eskimo Kitchen Mechanic," Berstead Mfg. Co., Fostoria, OH.

Domed chrome motor has single shaft, natural handle and off/on knob. Stands 12″ and lifts off white metal, stepped base which has receptacle for reeded, tapered clear glass. Lift off feature makes it portable.

$30 (with original glass).

Mary Dunbar "Handymix" Mixer: 1930s. Chicago Electric Mfg. Co.

Stands 11½″. White metal with removable motor for "portable" use. Push button off/on switch and no speed control. White wooden handle and tiny beaters. You need to be strong to use this one at the stove! $20

Chronmaster "Mixall": Mid 1930s. Chronmaster Electric Corp., New York & Chicago.

Stands 14″ and has chrome and black painted motor on black painted base. Off/on switch on front. Chrome knob on top is attached to single shaft which raises completely out of the way for glass removal or insertion. Original silver banded glass fits hole in base.

$30 (with original glass).

Dominion "Modern Mode" Mixer: 1932–1933. Dominion Electrical Mfg. Co., Minneapolis, MN.

Faceted, angular Art Deco body and base with top, rear 3-speed lever control. Runs on AC or DC and has 2 custard glass bowls and juicer. Also has a mechanism that controls beater height. **$75**

General Electric Mixer: 1938. G.E. Corp.

Ser. #10-A. Art Deco design has upright housed motor with top speed control. Note three beaters in a row. Work light next to beaters shines into bowl. Two white glass bowls and black Bakelite handle. This was my grandmother's! **$40**

Hamilton Beach Mixer: 1930s. Hamilton Beach, Racine, WI.

Model "G." Cream metal with black Bakelite handle. Off/on lever controls "Mix Guide" in window below handle. Bowl control lever in base. Mixer lifts off to be portable. Two white glass bowls. **$35**

Hamilton Beach Malt Machine: Mid 1920s. Hamilton Beach Builders, Racine, WI.

The forerunner to the home malted maker. The "cyclone #1" was the first in 1924. Stands 19″ and is run by the "home motor" introduced in 1912. Heavy nickel housing and square stand which houses interior spring operated switch. White marble base. **$175**

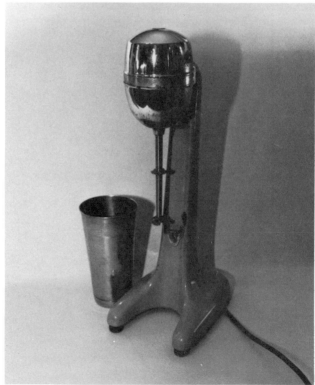

Hamilton Beach Malt Machine: 1930s. Division, Scovill, Racine, WI.

Standard of the industry stands 18½″. Appropriate in home or business, the gleaming heavy chrome motor with rear switch is mounted on green enameled, cast iron base. A "push up" switch is activated when the stainless steel cup is in position. Sometimes difficult to obtain the correct cup.
$75; **$85** (with cup).

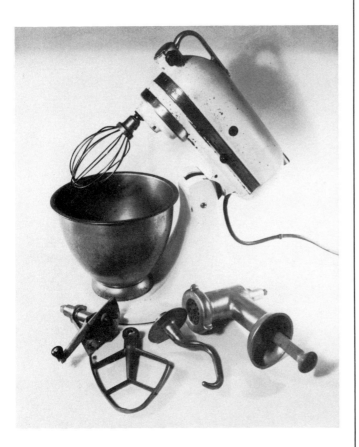

Hobart KitchenAid Mixer: 1939. Hobart Corp., Troy, OH.

Model K 4-B. Looks much like the KitchenAid mixers made today but it is quite a bit heavier being made entirely of metal. Cream painted body trimmed in heavy aluminum. Heavy cast aluminum bowl screws down to base. Also has beater, whisk, can opener, meat grinder & dough hook.
    $50 (mixer only); **$8** (each additional attachment).

Kenmore Hand Mixer: 1940s. Sears, Roebuck & Co., Chicago, IL.

Cream colored plastic hand mixer has one $4\frac{1}{2}''$ beater and advertises that it beats, whips, stirs, mixes and mashes! Not a bad trick for a one-speed mixer! Original box with booklet, hanger plate, and original warranty.
    $25; **$15** (mixer only).

Handy Hanna Mixer: 1940s. Standard Products, Whitman, MA. Knapp Monarch, St. Louis, MO.

White metal, dome top with toggle switch and varnished natural handle. What makes this interesting is that it screws onto a plain quart jar. Paper label on motor.

$18

Sunbeam "Mixmaster": 1930s. Chicago Flexible Shaft Co., Chicago, IL.

Rotating handle allows for juicer and other attachments. Rotating rear "bullet" speed control and large Sunbeam decal on side. Black & white. Trigger mechanism to remove mixer from base. Shown here with two bowls and original booklet.

$35, **$40** (with original booklet).

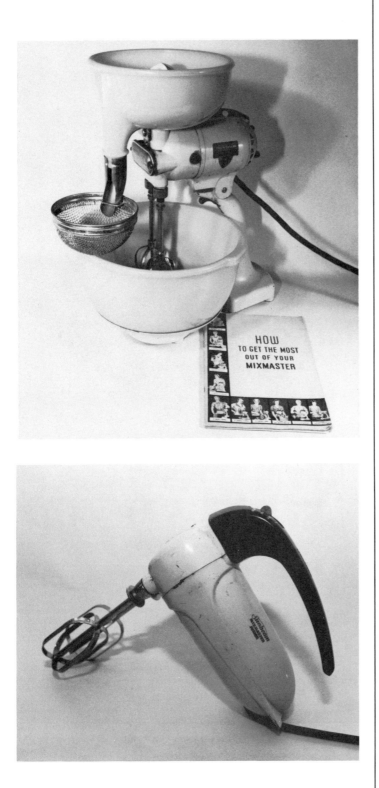

Sunbeam "Mixmaster": Early 1930s. Chicago Flexible Shaft Co.

Model "K". Cream colored body with fold-over black wooden handle, rear speed control and light green, opaque Depression glass bowls. Juicer attachment with wire strainer. Later models had strainer that fit inside juicer. Original booklet.

$65 (with two green bowls);
$5 (original booklet).

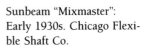

Sunbeam "Mixmaster Junior": 1940s. Sunbeam Corp., Chicago, IL.

Torpedo shape in white with red speed control on black plastic handle. Probably the forerunner of today's truly portable mixers. No stand and no bowls. **$15**

Sunbeam "Mixmaster":
1940s. Sunbeam Corp.,
Chicago, IL.

White and black with bullet-
like control dial on rear.
Beater ejector and bowl size
control knob in base. Juicer
has aluminum strainer inside
bowl. Original booklet.
$35 (mixer only);
$5 (booklet).

Inside cover of a Sunbeam
"Mixmaster" booklet show-
ing the various attachments
designed to lessen the
drudgery of kitchen tasks.
These were made from the
mid 1930s into the 1950s.
Some of the line was
dropped soon after World
War II.

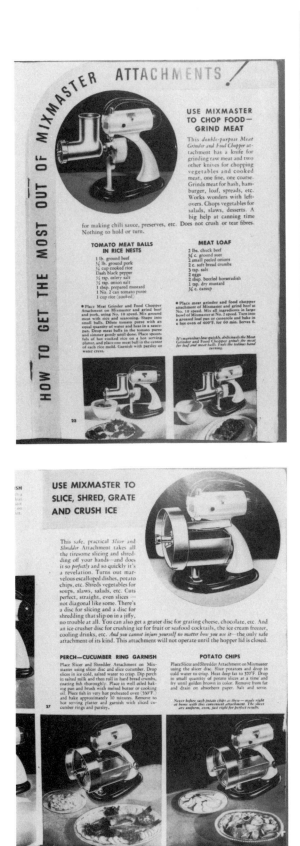

This shows the food chopper/meat grinder attachment. Please note that this is driven by the power unit which is required to operate several of the attachments shown. It mounts between motor and fits snugly into two holes in base.

$8 (power unit), **$8** (grinder/chopper).

Used to slice, shred, grate and crush ice. Came with slicer disk, shredder disk and ice crusher disk. Note mixer mounted directly to slicer/shredder attachment and required no power unit.
$25 (grater, slicer, shredder with three blades).

Used to slice beans. This mounted directly to the "Mixmaster" and served as its own power unit. Simply attach to beater connections and place in bowl. Feed beans into elevated slot in disk top. **$20**

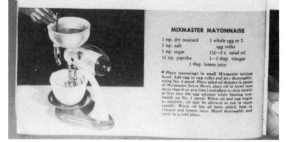

Juicer/mayonnaise maker. Consists of glass juicer bowl, metal spout, strainer (later models had interior strainer) and the elusive oil dripper. **$8** (complete).

Potato/apple peeler. "It's almost human! Goes over the bumps—into the hollows." Advertised no waste and so efficient. "No mess, no water, no more chapped hands." **$20**

Drink Mixer: Advertised "This is a high-speed drink mixer, just like the fountain mixers, and is many times faster than the Mixmaster itself. Complete with large glass mixing vessel." **$15** (drink mixer attachment with glass), **$25** (coffee grinder. Required drive unit—used to grind cereal, wheat, rice, etc.).

Can Opener: Requires power unit. Suggests: "... especially useful for opening cans in which the contents have been heated before the can is opened." **$5**; Knife and Scissor Sharpener: Requires power unit. "It sharpens your knives and scissors in less time than it takes to tell it." **$8**.

Polisher and Buffer: "A soft, high-speed polisher and buffer that gives a beautiful bright luster to all your silver pieces." Came with generous supply of polishing compound. **$2**; Freezer Attachment: "Does away with the one disagreeable part of making ice cream at home—the tedious, tiresome hand-turning." **$8**

USE MIXMASTER TO SHELL PEAS TWICE AS FAST *much Easier*

Until you see the way this marvelous *Pea-Sheller* attachment pops the peas out of the pods, twice as fast and much easier than ever possible by hand, you have no idea how simple this task can be. All you do is feed the fresh peas into the sheller. The empty hulls drop out on the other side — the peas DROP INTO THE BOWL. Big pods or small it makes no difference. This marvel shells all sizes and shapes and never misses a pea. And FAST! — it shells peas just as fast as you pick them up, singly or two at a time. Saves time. Saves your fingers. Causes no muss. In fact, it beats any other way of shelling peas all hollow. It sure does a BIG job quickly and *well*. Goes on your Mixmaster in a jiffy.

USE MIXMASTER TO CHURN BUTTER

With this new, fast and efficient MIXMASTER BUTTER CHURN attachment, Mixmaster does the arm-work of butter making for you. Saves time, too. Enables you to make your own butter, either salted or unsalted to suit your taste, easily and quickly. Same procedure as usual *less the work*. Built to last with stainless steel, rust-proof inside parts, wood paddles, sturdy 4-quart crystal glass churning jar.

Pea sheller: "All you do is feed the fresh peas into the sheller. The empty hulls drop out the other side—the peas drop into the bowl." Requires the power unit. **$8**; Churn: "Same procedure as usual less the work. Built to last with stainless steel, rust-proof inside parts, wood paddles, sturdy 4-qt. crystal glass churning jar." **$35**

Use **MIXMASTER**

To Rice Potatoes; Pureé Peas, Tomatoes, Beans, Spinach, Prunes, Apricots, etc.; To Cream Bananas; For Raspberries, etc., cooked or fresh; For Jellies, etc.

For delicious purees of all kinds, for tomato juice, jellies, banana pies and cake fillings, cream of vegetable soups, riced potatoes, fresh or cooked berries, velvet-smooth apple sauce, etc., this marvelous *Colander* attachment for your Mixmaster has no equal at any price. It takes no end of tiresome difficult work off your hands. *The complete answer to the doctor's recommended way to prepare those highly necessary vegetables— spinach, peas, etc.—for babies and small children.* Saves tedious, painstaking hand-straining. Saves buying specially prepared canned vegetables for the little ones in expensive half-pint tins. It's simple to operate—and FAST. Drop vegetables or fruit into the big, generous-sized hopper and touch the switch. All done in a jiffy with results beyond comparison with hand or other methods. Quickly attached, easy-to-use, strong, sturdy, substantial.

Ricer: Use it to rice potatoes, puree peas, tomatoes, beans, spinach, prunes, apricots, etc.; to cream bananas and for raspberries, etc., cooked or fresh; for jellies, etc. Requires use of power unit. **$20**

Sunbeam cabinet for Automatic Mixmaster and attachments: "The cabinet plugs into your regular wall outlet. There is a conveniently located double outlet on the side of the cabinet itself so two appliances can be operated at the same time if desired." Cabinet stands 60½″ and is 24″ wide. Has sanitary porcelain table top that slides out to give a working space and out of the way when not in use.   **$175**

This New Handsome *Sunbeam* Cabinet
for Your *Automatic* MIXMASTER *and Attachments*

Just what you want for keeping your Automatic Mixmaster and its attachments all together—right at your fingertips whenever you need them. The cabinet plugs into your regular wall outlet. There is a conveniently located double outlet on the side of the cabinet itself so two appliances can be operated at the same time if desired.

Has lots of space above and below for cooking and baking equipment, and a porcelain draw out table top to work on. There's an individual compartment for your Automatic Mixmaster itself, and 5 others for attachments behind the upper doors. There's a roomy drawer below the table top for cutlery, etc., and generous cupboard space beneath that with a convenient draw-out shelf for more attachments, pans, etc. Also a handy door-rack for pot covers.

PORCELAIN DRAW-OUT TABLE TOP

The sanitary porcelain table top slides out to give you a convenient working space (24x20 inches) and slides back out of the way when not needed.

The cabinet and table-top are finished in the lovely ivory of the Automatic Mixmaster itself, with fine hardware. Attractive rich black base of modern design. 60½ inches high, 24 inches wide and 20 inches deep when table top is in. $24.75 (Denver and West, $26.75). Freight prepaid.

"Made-Rite" Drink Mixer: 1930s. Weinig "Made-Rite" Co., Cleveland, OH.

Light weight mixer in cream and green painted metal has no switch and single mixing shaft. Name is on decal on front of backpiece/base where a shaped glass holder is permanently mounted.   **$20**

Dorby Whipper: 1940s.

Model E. Chrome motor housing has black Bakelite handle and off/on switch. Clear glass Vidrio, measured bottom.   **$25**

Electromix Whipper. Chicago, IL.

Ivory colored, offset metal motor housing has push-down break mechanism. Filler hole in lid. Measured, clear glass base. Stands $7\frac{1}{2}''$.

$25

Knapp Monarch "Whipper": 1930s. Knapp Monarch, Belleville, IL.

Chrome-housed motor, single shaft, black wooden handle. Off/on switch on lid. Top rests on a flat tripod over 3-cup clear measured base. 7″ tall.   **$20**

**Knapp Monarch Whipper:
Mid 1930s. Knapp Mon-
arch, St. Louis, MO.**

Stands 9½″ from top of red
plastic handle. Motor housing is white painted metal
and has red off/on knob.
Shaft has a white plastic
beater. Base is of milk glass
and has reeded, fin feet.
Good styling.   **$25**

**High-Speed Whipper: Early
1930s. Knapp Monarch
"Moderne." Knapp Mon-
arch, St. Louis, MO.**

Whipper has a 3-cup clear
glass bottom with top motor
housing in cream colored
metal and Depression green
open handle. A paper label
decorates front. Stands 8″
overall. Just plug it in and
stand back! No switch.   **$20**

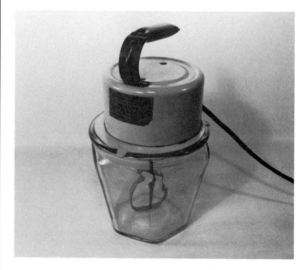

**Kenmore Whipper: 1940s.
Sears, Roebuck & Co., Chicago, IL.**

Cream metal domed top sur-
mounted by dark blue Bakelite knob. No switch. Clear
glass bottom has Art Deco
styling with vertical fins
forming feet. Stands 8½″. **$15**

Unmarked Whipper: Late 1920s–Early 1930s.

Stands 7½″ with green motor housing and green Depression glass cup. Cup is embossed on bottom "Vidrio Products Corp. Cicero, Ill., Made in U.S.A. #E 20." Note unusual serpent shape of mixer shaft. **$20**

Unmarked Whipper: 1930s.

Offset Depression green metal motor housing. Has filler hole in lid. Nickel handle. Measured base is marked "Vidrio" and is green Depression glass. **$25**

Unmarked Whipper: 1930s.

Offset Depression green metal motor housing. Has filler hole in lid. Nickel handle and brake mechanism. Clear, measured glass base. Even the cord is green. **$20**

Step off the
Endless Treadmill
of KITCHEN WORK

... *The hard, tedious tasks of stir-ring, mixing, beating, whipping, etc., necessary to food preparation* ... *with the help of* Gem Kitchen Mechanic

In 1929 the "Kitchen Mechanic" appeared in this ad. Note similarity to a KitchenAid mixer. This one probably died as a result of the crash of '29.

CLICK A SWITCH...
and a POLAR CUB goes to work

In many thousand homes these nimble aides speed up household routine. Polar Cubs are Christmas gifts that *do things!* Electrical devices built to lift the more tedious little chores right out of housework.

Click the switch of the new all-purpose Polar Cub Beater, and it mixes light batters and mayonnaise, whips cream, beats eggs, to a fluffy lightness, in jigtime.

When you want fresh orange, lemon or grapefruit juice, the Sunkist Junior Juice Extractor swiftly yields every drop without a bit of muss or labor. Mix drinks, salad dressings and other light mixtures with the Polar Cub Mixer, which also whips cream and eggs. Indoor drying, whether hair, lace, wet shoes or clothing, is the job for the Polar Cub Dryer. These are only a few of the modern Polar Cubs.

The heart of every Polar Cub is the triple-tested Polar Cub electric motor, fully guaranteed by its makers.

Look them over on this page. Aren't there *ideal* gifts here for certain people on that Christmas list? Write their names on the margin of the page. Use it as a shopping guide. You'll find Polar Cubs at any modern store. If not, you may mail your order on the coupon, to The A. C. Gilbert Company, 138 Erector Square, New Haven, Connecticut. Or send for Free Booklet.

$11.95
POLAR CUB BEATER
Made Expressly for
THE WESSON OIL-SNOWDRIFT PEOPLE

Mixes mayonnaise and all light batters. Whips cream, beats eggs. It does the job thoroughly and quickly. Its powerful motor and double intertwining agitator aerates evenly. Its standard rod is *curved* to enable the use of large mixing bowls. The blades are easily detached for washing.

• • •

SUNKIST JUNIOR JUICE EXTRACTOR
Adopted by the
CALIFORNIA FRUIT GROWERS EXCHANGE

It provides fresh orange, lemon or grapefruit juice as quickly and easily as turning on a faucet. Press the fruit gently down at the top, and every drop of juice gushes from spout. Easy to clean.

$4.95 and **$7.50** for larger size

POLAR CUB MIXER. Just the thing for malted milk, chocolate and other drinks and light mixtures with soda fountain speed and efficiency.

$14.95

Polar Cubs

The A. C. Gilbert Company, 138 Erector Square, New Haven, Conn.

———Send me "The Book of Polar Cubs—Electrical Helpers for Every Household."
———My dealer cannot supply me with Polar Cub.
Send me _____

I Enclose $ _____
Name _____
Address _____

$4.95

POLAR CUB HAIR DRYER. There are many uses for it about the house, including drying hair, wet shoes, clothing, laces, etc. It blows a warm, dry current of air.

December 1929 Good Housekeeping

In December 1929, the A.C. Gilbert Co. (Erector Sets) made the Polar Cub Mixers expressly for the Wesson Oil Snowdrift Co. (also made the Sunkist Juice Extractor.)

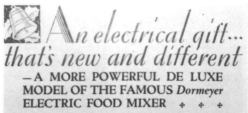

The 1929 Dormeyer mixer is pictured in an ad from *Good Housekeeping*. It would stand up by itself.

Hamilton Beach showed off the attachments for their latest Food Mixer in this original booklet from the 1930s.

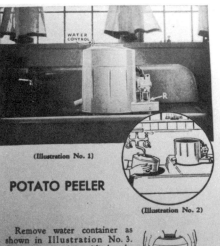

(Illustration No. 1)

(Illustration No. 2)

POTATO PEELER

Remove water container as shown in Illustration No. 3. Grasp top container with thumbs and press down on the metal rim with your finger tips. Be careful not to wedge top container by lifting one side higher than the other; lift up evenly and top container will come off easily. Fill with cold water, see Illustration No. 2.

(Illustration No. 3)

Lift the motor from the Food Mixer and place it on Peeler as in Illustration No. 1. Press motor down until square shaft "R" engages the coupling on the motor; to assist, turn plate in bottom of Peeler back and forth slightly.

[30]

Hamilton Beach Food Mixer

(Illustration No. 1)
Slicing

(Illustration No. 2)
Shredding

(Illustration No. 3)

SLICER and SHREDDER

Remove the motor from the mixing stand and place it on the Power Unit. Press motor down until the square shaft "R" engages the coupling on the Mixer. Twist motor back and forth slightly if necessary to make the coupling slip over the square shaft. (This Power Unit is the same as is used with

[38]

Hamilton Beach Food Mixer

This Universal mixer from the mid 1930s featured a detachable motor for portable use. They were giving away free mechanical juicers that year but you could also purchase an electric juicer or vegetable slicer by that time.

This late 1930s "Mixmaster" ad featured the new "Automatic Mix-Finder," a bullet-like affair on the back.

This 1934 *Better Homes & Gardens* ad showed a revolutionary new G.E. Hotpoint mixer that had the motor in the base instead of on top. "That's important! Why? Because no oil can drop into your food in the mixing bowl."

Mixmaster boasted 10 speeds and 60% more power with the new full-mix beaters in 1935.

A 1935 ad showing an array
of Universal appliances.

In September of 1935, this
General Electric Hotpoint
mixer featured stationary or
portable use anywhere in the
kitchen.

Fitzgerald's "Magic Maid" mixer in 1935 said it could do more than the rest with the aid of attachments. It had all of the accessories that others had, but it could even roast coffee and pop corn. I would love to see these attachments!

$25 (mixer only, current value); $50 (with "normal" attachments); $95 (if you can roast or pop).

Hamilton Beach advertised one hand operation in this 1935 ad and introduced a unit that could be used separate from the base.

SENSATIONAL NEW KITCHENAID
that's TWICE as EASY to OWN!

The Food Mixer that "Does It All"...at an Exciting Price!

POWERFUL AND PRACTICAL FOR ALL HOUSE-HOLD MIXING, BEATING, WHIPPING, KNEAD-ING, CHOPPING, SLICING, GRINDING, SHRED-DING, SIEVING AND ... MANY OTHER TASKS

The New Model "K" . . . beautiful modern lines and gleaming white Dulux finish trimmed in chromium — a KitchenAid in ALL that KitchenAid means! Compact . . . Generous bowl capacity.

■ NOW WITH THIS splendid new KitchenAid—the *completeness* of KitchenAid food preparing service is available *as never before* to *every* American home! Perhaps you've dreamed of a KitchenAid—but never dreamed it *could* be so easy to own! Such a machine as this is the happy realization of our plans to bring *you* KitchenAid at a greatly reduced price.

Here's a KitchenAid styled by one of America's foremost designers — a model built and powered to "*do it all*" for you in the same masterly way as the larger Model "G" shown below. Full planetary action in mixing and beating, with *stationary bowl.* Three definitely selected, *constant* speeds. *Full* power at each speed. Practical attachments — see list.

The modest price may be paid on the Budget Plan — KitchenAid can be yours now, so easily. Let us send you *all* the good news.

FRESHLY GROUND COFFEE

The last word on the much advertised, much discussed subject of Coffee is Coffee FRESHLY GROUND for each meal, in the COR-RECT GRIND for your maker. This smart, little, low priced electric mill is always ready.

Attachments

Vegetable Slicer
Shredder Plates
Food Chopper
Colander & Sieve
Juice Extractor
Pastry Knife
Coffee Grinder
Pea Sheller
Knife Sharpener
and others

. . . Famous Model "G," the larger size

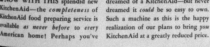

KitchenAid REG. U.S. PAT. OFF.

In 1937, KitchenAid intro-duced the "G" model, slightly smaller than the "K" and with a myriad of attach-ments and offered a "Budget Plan." Ad does not say just how much more economical it was!

Let Santa come early with **this** gift

Make CHRISTMAS DINNER Merry!

● **Open** *this* **gift before Christmas!** Let it make your preparation of extra-good things for the holidays *thrillingly easy.* With its exclusive "Thoro-Mix" action, and its extra power, this new KitchenAid Mixer helps you put supreme quality in foods: fruit cake, plum pudding, cookies, cakes, candies, rolls, cran-berry sauce, mince meat, fresh pumpkin—and soups, salads, vege-table and meat dishes. It operates labor-saving food preparing attach-ments without any extra "power adapter." See it at your dealer's now—or write KitchenAid Div., The Hobart Mfg. Co., 412 Penn, Troy, O.

MODEL 3A
KITCHENAID $29.95 IN U.S.A. WITH JUICER

This model 3A KitchenAid mixer ad in 1940 urged to open this gift before Christ-mas.   **$50** (current value).

101

Blenders were still just a little new when this November 1940 ad appeared for the Universal "Mixablend."

$15 (current value).

Sunbeam ad of 1943 "... making rationed foods go further." This ad also stated that "There have been no Mixmasters made at the Sunbeam factory since Spring, 1942. Production of war goods replaced them at that time. But they will be back with Victory. Get yours with a War Bond."

Novelty Appliances

■

W hen trying to classify objects of any kind, there always seems to be those articles that just don't fit into any prescribed category. This chapter is devoted to those unique items that fall into this group. While this is not meant to be an exhaustive listing of every offbeat appliance ever manufactured, it will give you some idea of what to look out for.

Appliances such as the electric flour sifter, coffee grinders, knife sharpeners, and electric tea kettle will appear here. Because it was used for so many different functions, the Universal home motor could be listed under Blenders, Mixers, and Whippers or Combination Appliances. However, since it was a unique approach to running otherwise manual appliances, we have listed it here.

While not exactly an appliance in the regular sense of the word, we have included a Montgomery Ward timer. The clock is of the wind-up type but there is a cord in the back that plugs into a wall outlet. It has a receptacle in the back for an appliance to plug into. This is an odd one, but that's what makes these novelty appliances interesting.

7

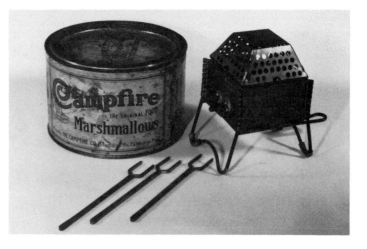

"Angelus-Campfire Bar-B-Q Marshmallow Toaster": 1920s. Campfire, Milwaukee, WI.

Measures 3″ square with flat-topped, pierced pyramid top piece. Base stands on loop wire legs with rubber encased "feet." Also has flat wire forks for roasting marshmallows. These are a little hard to obtain. **$55**

"Universal" Home Motor: 1917. Hamilton Beach Mfg. Co., Racine, WI.

High-speed motor runs on AC or DC and could be adapted to do several tasks from running a treadle sewing machine to turning an ice cream freezer. This one also has foot pedal control (not shown) set up for a sewing machine. This was the first, high-speed motor. It also powered the first malt machines.
     $30 (no attachments).

Coffee Grinder: 1936. "KitchenAid." Hobart, Troy, OH.

Model #A-9. Heavy, cream colored cast base houses motor and has coarse/fine adjustment at neck. Clear glass jar container has screw-off top and served as storage for coffee beans. Simple off/on switch. The first home coffee grinder. **$60**

"Vita-Juicer": 1930s. Kold King Dist. Corp., Los Angeles; Hoek Rotor Mfg. Co., Reseda, CA.

Stands 10″ and is of very heavy, cream-painted cast metal. In three parts with base motor, container and top. Fitted lid with lock groove, lock down wire and nickel handle. An aluminum pusher fits hole in top. **$35**

"Miracle" Flour Sifter: Mid 1930s. Miracle Products, Chicago, IL.

"Nothing new under the sun"—this electric sifter was introduced circa 1934. It has cream body and blue painted wooden handle, on the base of which is "hold down" on button. It really works as well as the more recent ones. Vibrates flour through wire strainer. **$35**

Tea Kettle: 1930s. Mirro Aluminum.

Four qt. model has flat base with heating element inside. Nickel chrome domed body has spout with "whistle." A large Bakelite handle adorns one side. **$20**

Clock/Timer: Late 1930s. Manufactured for Montgomery Ward & Co.

Works beautifully but strange in that the clock is wind-up while the cord running from the back plugs into the wall. It has a receptacle for an appliance. Body is cream with silver and red face. Curved glass. Body swivels on weighted base. Innovative for the time. **$25**

Tea Kettle: 1910s. Universal (Landers, Frary & Clark), New Britain, CT.

Nickel-plated body and base in one piece (#E973). Rather squat body but "cute." High curved handle is black painted wood. These are not very prevalent. **$45**

Landers, Frary & Clark urged women to "Universal-ize" their homes in this 1921 *Good Housekeeping* ad.

The Hamilton Beach "Home Motor," as seen in this ad of 1926, needs no screws or bolts to attach the motor.

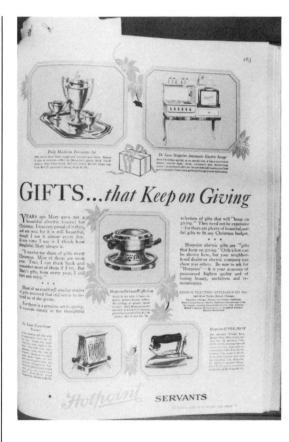

A 1927 Hotpoint Christmas ad.

A 1928 Christmas ad for Hotpoint appliances. It features the "Florentine" coffee set, "De Luxe" waffle iron, Hotpoint's first "Automatic" toaster and others.

An early ad for the "El" includes nearly every line of appliances made by Hotpoint.

This KitchenAid model A-10 was the first home coffee grinder.　　**$60**

An early Universal ad stated that servants were easier to keep when you had Universal appliances.

Toasters

■

When ordinary bread is toasted, something wonderful happens. This great American breakfast staple has been around since the time of the ancient Egyptians. They are said to have scorched bread to remove the moisture as a form of preservation. It is noted that the Romans brought the custom to Britain. By the seventeenth and eighteenth centuries, toasting had become common practice with the British as well as the Americans.

The quality of early colonial toast is somewhat doubtful. At first, locally smithed, crude, iron fork devices were made to hold the bread over a smoking fire. The first manufactured toasters in this country were the tin and wire contraptions designed to set over a coal stove or on a gas burner. These were made as early as 1819, and are again being reproduced.

General Electric and Westinghouse ran neck and neck to develop and market the first electric toaster. G.E. won. 1905 saw the unsuccessful model X2—a wire device without body or shelf. In 1909 a successful toaster, model D12 with a white porcelain base, was introduced. In June of 1910 Westinghouse ran a headline in *The Saturday Evening Post:* "Breakfast without going into the kitchen—ready for service any hour of the day or night."

It was not uncommon for manufacturers to bring out dual-utility items, so the idea of an appliance designed for just one function was still a bit strange. The Armstrong Electric and Manufacturing Company of Huntington, West Virginia, offered a three-in-one combination grill, hotplate, and toaster. It retailed for $8.95.

The next two decades witnessed considerable innovation and progress as manufacturers mastered the handling of electricity for household tasks. In 1922 the Estate Stove Company of Hamilton, Ohio, came out with an ingeniously designed four-slice toaster on which "one movement of the lever knob turns over all four racks." "The newest turnover-toaster, a Hotpoint servant" was introduced the same year by the Edison Electric Appliance Company. The popular sandwich toasters now appeared on the market. One of the first was the "Hostess" made by the All Rite Company. Designers vied with unusual

8

shapes and concepts for toasters. In 1922 a Detroit firm, the Best Stove and Stamping Company, came out with perhaps the first "oven type" toaster with an in-and-out sliding bread rack, an interesting idea which Landers, Frary & Clark "used" the following year.

The most significant development was the automatic toaster. That innovation is credited to D.A. Rogers who, in 1924 (two years before Toastmaster), designed a toaster equipped with a dial somewhat like that on a telephone. By setting the dial to the desired type of toast, the toaster operated until it automatically shut off, at which point the hinged sides would drop down so the toast could be removed. Little else is known about this device.

In 1919, Glen Waters was experimenting with a spring, motor and switch for a toaster. He was fed up with burned toast. He custom built a few four slice toasters for restaurant use. Most were returned, but only minor adjustments were required for the idea to work. Waters, with his backer Harold Genter, formed the Waters Genter Company in 1921. By 1925 the name "Toastmaster" was registered. Murray Ireland joined the company and designed Model 1-A-1, the first household automatic toaster. It was marketed in 1926 and launched the company into the housewares field making toaster history. No one knows for sure who came out with the first "flip-flop" toaster. These appeared in the 1920s and remained very popular until after World War II.

Innovations in the toaster industry were as varied as in any other appliance. As early as 1930, Proctor brought out a model that operated on the basis of the surface temperature of the bread. In 1935 Knapp Monarch announced its "Tel-A-Matic" (model 5140) toaster which "remembered" to allow more time for a cold toaster and less time for a very hot one. Eventually the tic-toc mechanisms were replaced with silent ones. There was a combination coffee pot and toaster, hotplate and toaster, automatic flip-flops, and a myriad of other designs. One, the Toast-O-Lator (1937) used a conveyor device that took the bread in a slot at one end and passed it through, plopping it out at the other end. As early as 1940, G.E., again with a "first," introduced a "keep warm" feature that after the heater was turned off, the toast could be held down inside the toaster, where residual heat would keep it warm until it was wanted.

Today General Electric's small appliance division is owned by Black & Decker. It boggles the mind to think what this could mean to toast!

Dominion Toaster: Mid 1920s. Dominion Mfg. Co., Minneapolis, MN.

Bright pierced chrome body (this one never used) with green wooden "pick me up" handles, detachable cord. Small Bakelite tab door openers.

$30 (mint condition).

Edison, Hotpoint General Electric Toaster: Late 1910s–1920s.

Cat. #156T25. Flip-flop type with pierced nickel body. Don't be too confused by the name. This one came in several models, names etc. Biggest difference was metal or knob handles. Identical mechanisms. **$20**

Edison Toaster: Mid to late 1910s. Edison Appliance Co., NY.

Cat. #214-T-5. Nickel open body with free swinging tab closures at top. Has single knob and wonderful little removable toast warming rack. Detachable cord. **$45**

119

General Electric Toaster.

Model D 12. Patented 1908
and generally thought to be
the first successfully marketed electric toaster. (3
years later than the experimental X2). Porcelain base,
wire body and removable
toast rack. Porcelain plug
screwed into socket light.
Also came decorated with
rose garlands.

$150 (complete with rack
and cord); $175 (complete,
decorated).

**General Mills Toaster: Early
1940s. Minneapolis, MN.**

Cat. #GM 5 A. This 2 slice
pop-up in chrome with
wheat decoration has sym-
metrical Bakelite support en-
compassing body. Operates
on AC or DC. Red knob
serves as light/dark control.
Has opening crumb tray.
Great design.   **$20**

Handy Hot Toaster: 1935.

Type EAUK. Flip-flop in
chrome body, embossed
reeded center decoration,
natural wooden handles, tiny
reeded feet.   **$15**

Heat Master Toaster: 1923–35.

Square chrome body with rounded corners, end opening, 2 slice manual operation. Black Bakelite handle and feet. Bottom opens for crumb removal. Good design but your bread would fall off rails! **$30**

Kenmore Toaster: Early 1940s.

Mechanical 2 slice pop-up toaster in chrome body. This has rounded edges and black Bakelite handles. Mechanical clock mechanism and light/dark control. Detachable cord. Design on body sides. **$15**

Knapp Monarch Toaster "Reverso": 1930.

Cat. #505. Little rectangular nickel body with top rounded corners on black painted base. Flip-flop pierced doors have open tab handles. Poor little thing doesn't even have mica to separate wires that are just stretched from here to there. **$15**

Knapp Monarch Toaster: Mid 1930s.

Cat. #21-501. Chrome, rounded body, flip-flop type. One handle opens doors individually. Art Deco design on doors, Bakelite handles. Interesting opening mechanism.   **$20**

Manning-Bowman Toaster: Early 1920s.

Open nickel body. Black Bakelite handles open toast "cages" that turn completely over. Detachable cord. This one is in rough condition.
         $55 (mint condition).

Manning-Bowman Toaster: 1920s. Meriden, CT.

Cat. #81. Flip-flop type with chrome body, stepped Bakelite feet and round knobs. Doors have reeded decoration. Detachable cord.
                     $15

Merit Made Toaster: 1930s. Merit Made, Inc., Buffalo, NY.

Model "Z". Round, silver painted body on black base. We call this flip-flop "Buck Rogers." Plunger on top opens both sides simultaneously. Attached cord, Bakelite handles.   **$30**

Montgomery Ward & Co. Toaster: Mid 1930s.

Model #94-KW 2298-B. Flip-flop type with brown Bakelite handles. One handle opens both doors simultaneously. Pressed design on door (removable cord).   **$18**

Omega Toaster: Early 1930s.

Low rectangular nickel body with spring loaded doors that bulge at bottom. Red Bakelite handles and feet.
          $20

Porcelier Toaster: 1930s.
Porcelier Mfg. Co., Greensburg, PA.

Most unusual with basket
weave porcelain body and
floral transfer decoration.
Clock mechanism pop-up
and porcelain knob
light/dark control. This is
part of a large breakfast set.

$75

Sun-Chief Toaster: 1930s.
Sun Chief Electronics,
Winsted, CT.

Series 680. Manual flip-flop
in angular chrome body with
reeded Art Deco door and
top decoration. Doors have
vented bottom. Black
wooden handles.   **$15**

Steel Craft Toaster: Late
1920s.

Don't know much about this
one. Flip-flop type with
green wire open body. Handles and feet are red painted
wood knobs. Should be
guaranteed to burn your fingers. Attached cord.   **$35**

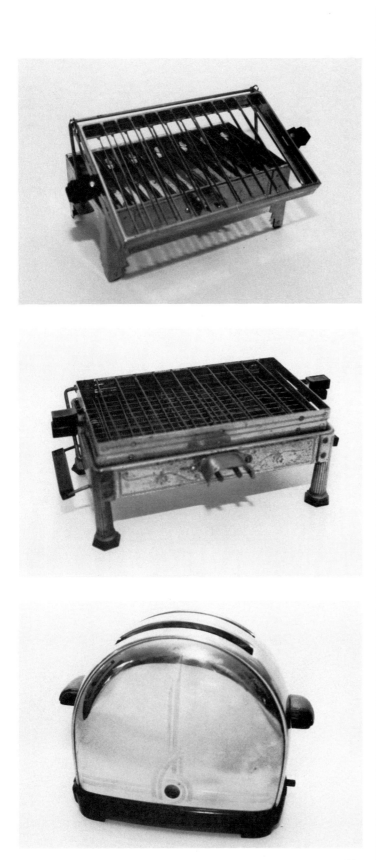

Sunbeam Toaster: Early 1920s.

Model 4. 5″ × 9″ flat rectangular chrome, decorated body on angular Art Deco flat legs, tab feet. Double wire holder with manual flip-flop operation. This does not have the drop handles.
$45

Sunbeam Toaster: Early 1920s.

Model B. 5″ × 9″ flat rectangular chrome, decorated body on reeded legs and hexagonal Bakelite feet. Double wire holder with manual flip-flop operation. Little drop handles on sides for lifting while hot. **$45**

Sunbeam Toaster: 1932–42.

Model #T 9. Rounded chrome Art Deco Body, black Bakelite base and handles. This "Cadillac" is a dual slice automatic pop-up with Art Deco design surrounding indicator light, dual thermostats and voltage regulator. Crumb removal tray. Also came with rectangular liner with four glass trays. **$50** (toaster only); **$75** (toaster with liner and trays).

Sunbeam Toaster: 1936. Chicago Flexible Shaft Co.

Good Art Deco design in chrome with black Bakelite base and handles. Detachable cord in rear. Light heat indicator in front. Two slice automatic. Includes fitted, divided glass tray.　**$65**

Superlectric Toaster: Late 1930s. Superior Electric Prod. Corp., St. Louis, MO.

#66-4. Rounded chrome body with drop sides and wood handles. Wide enough to accomodate four slices. Operates on AC or DC.
　$18; $45 (never used and with original box, as shown).

Toastmaster Toaster: 1927. Waters-Genter Co., Minneapolis, MN.

Model 1A-1. Acclaimed as the first automatic pop-up toaster. This classic has louvered sides in the chrome body which is Art Deco in style with rounded front. Manual clock mechanism and light/dark control from A–G with panic button. **$80**

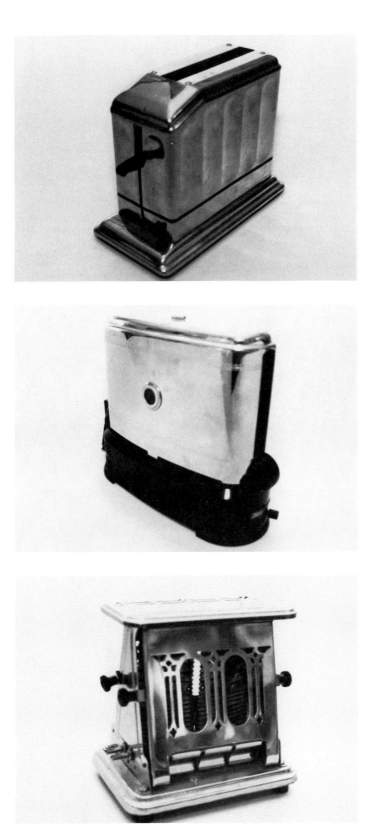

Toastmaster Toaster: 1929. Waters-Genter Co., Minneapolis, MN.

Model 1A-3. The third model Toastmaster in stylish chrome rectangular body. Vertically scalloped side decoration. Mechanical clock mechanism and light/dark knob. Attached cord. **$35**

Toast-O-Later Toaster: 1938. Crocker-Wheeler Elect. Mfg. Co. Model G or J (identical in appearance).

Mechanical wonder in chrome body and high Art Deco style having ovoid shape. Toast "walks" through on little tooth-like conveyers and falls out opposite end. Bakelite base & knobs. Adjustable control and on/off switch. Attached cord. Unique design. **$80**

Universal Toaster: Mid to late 1910s.

Flip-flop type with nice design on pierced doors and top. Black Bakelite handles and feet. Flat pierced warming rack, detachable cord. **$25**

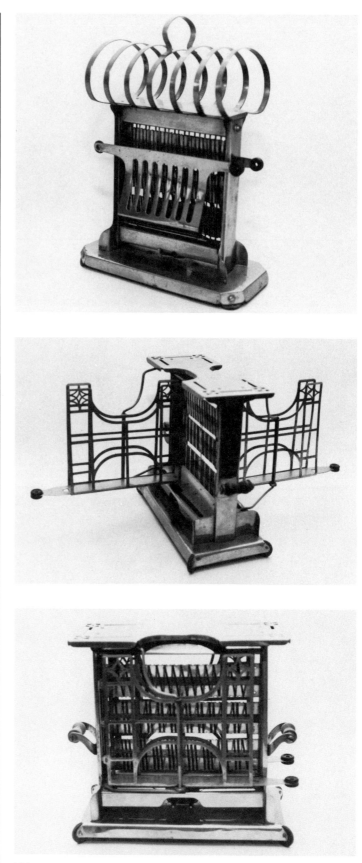

Universal Toaster: 1913–1920s.

Nickel body on flat base with tab feet. Pierced concave spring loaded doors and permanently attached warming rack. Good design.

$40

Universal Toaster: Late 1910s–1920s.

This could be called a "swinger." It has heavy nickel body on chromed steel base. Mechanism is really interesting. The vertical wire hinge in the center allows racks to swing out and around. Small wooden carrying handles, scooped out and pierced top warming rack.

$60

Universal Toaster: 1920s.

Model #E-3612. Flip-flop type in chrome body on black base. Oval embossed design on doors, pierced warming rack and black knobs.   **$20**

Universal Toaster: Mid 1920s.

Rectangular body with pull-out door. This one has missing base and has been painted.   **$5** (as pictured); **$25** (in good condition).

Universal Toaster: Late 1920s. (Landers).

Model #E 7542. Mechanical toaster with clock mechanism. Very similar to 2 slice model, this one slice has vertical rectangular body with circular embossed design, light/dark lever and "emergency" release.   **$50**

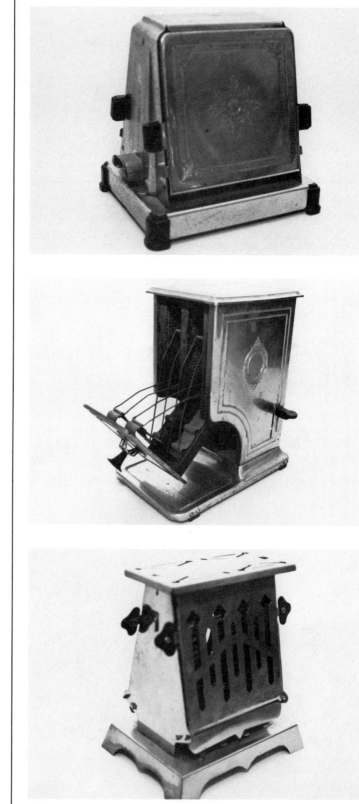

Universal Toaster: Early 1930s.

E221. Nickel body with light but nicely executed body designs. Flip-flop type with brown Bakelite handles and feet. **$20**

Universal Toaster: Circa 1930.

Model #E 7732. Mechanical toaster with clock mechanism. This large 2 slice has tall rectangular body with concave cut-out above base. Parallel line and embossed circular decoration. Light/dark lever and "emergency" release. Removable cord. **$75**

Unmarked Toaster: 1920s.

Nickel, pierced body with tab handles, cord, and flat warming rack. **$15**

Westinghouse Toaster:
1909. Mansfield, OH.

Type C. "Toaster Stove,"
Westinghouse's introduction
into the toaster market, ran
neck and neck with G.E.
This has flat, rectangular
body with 4 flat strip plates,
removable legs, tray, cooking
tray, wire rack, original box
and paper guarantee (never
used!).   **$125**

Westinghouse "Turnover
Toaster": Late 1910s. West-
inghouse Electric & Mfg.
Co.

Style #231570. Unusual in
that the element core is ce-
ramic rather than mica. Body
is nickel with open wire
doors. Black Bakelite knobs
on just one side. Note porcelain detachable plug.
Large, flat warming surface
on top.   **$35**

Westinghouse "Turnover"
Toaster: Mid 1920s. Mans-
field, OH.

Cat. #TT 3. Nickel body
with pierced doors and top.
Flat tab handles, detachable
cord.   **$15**

A New Hotpoint Toaster

Toasts two large slices. Highly nickeled.
Fibre feet. Equipped with cord switch for
convenient turning on and off.

$5.75

Toast the Hotpoint Way

Everyone likes toast. With a Hotpoint toaster you can make delicious,
brown toast right at the table and butter it while it's hot — ummmm!
You can't get enough of it. It's good for you, too.

Ask your dealer to show you his line of Hotpoint
toasters. *And take one home.* You'll find it will
give the whole family enough satisfaction in a
day to make it worth more than it cost — a daily
satisfaction that will be repeated for many years
to come, because every Hotpoint appliance has
quality that *endures.*

But be sure to ask for "Hotpoint" to be certain
of lasting satisfaction and economical years of use.

Hotpoint Triplex Grill

Here's another Hotpoint servant
you'll be proud of. It boils, broils,
fries, toasts, poaches — *three operations at once.* Convenient for cool
summer breakfasts, prepared right
at the table, for luncheons or for
delightful, tasty suppers—$13.50.

EDISON ELECTRIC APPLIANCE CO., Inc.
Chicago · Boston · New York · Atlanta · Cleveland
St. Louis · Ontario, Calif. · Salt Lake City
In Canada: Canadian General Electric Company, Ltd., Toronto

Hotpoint

SERVANTS

WORLD'S LARGEST MANUFACTURER OF HOUSEHOLD ELECTRIC HEATING APPLIANCES

THERE'S A HOTPOINT ELECTRIC RANGE FOR EVERY PURSE AND PURPOSE

This July 1925 Hotpoint
Servants ad shows a new
toaster as well as the "Triplex Grill." "Be sure to ask
for 'Hotpoint' to be certain
of lasting satisfaction and
economical years of use."

This 1927 ad shows the
Universal "Turn Easy
Toaster" that cooked toast
"Golden Brown in a minute."

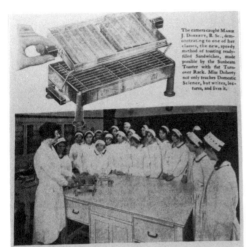

Girls Taught Advantages
of New Way Toasting

Favorite Delicacies Made 50% Quicker—Served Hot from the Toaster—More Delicious Than Ever

DOMESTIC Science is turning to a new and speedier way of toasting bread, of toasting crackers, plain or spread with food-paste or cheese, and of making Sandwiches "backwards"—that is, filling them *first*, then toasting. These ready-filled Sandwiches not only toast in half the time, but come hot from the toaster ready for serving. You've never tasted any so delicious.

All this is due to flat toasting principle, made practical by the Sunbeam Toaster's Level Turnover Rack. It grips the toast top and bottom. Hence, the filling can't spill out of Sandwiches even when you turn them. And crackers can't drop out of reach. Plain bread, too, is easiest turned this way.

Because the Sunbeam toasts two slices at once, directly over the heater where heat is most intense, toasting is 50% quicker. Hence, without waiting, more people are served. Breakfast consumes less time. The electric bill is lower.

Ask your dealer or Public Service Co. to show you the Sunbeam Toaster, $8.00. Send to us and we'll ship it direct, if you cannot find it nearby.

Manufactured and Guaranteed by CHICAGO FLEXIBLE SHAFT CO. 39 Years Making Quality Products, 5640 West Roosevelt Road, Chicago

Hinged Tray collects all crumbs. Snap it open and shut and the Sunbeam is cleaned. No recesses to gather refuse. Keeps the toaster sanitary.

Sunbeam
THE BEST ELECTRIC APPLIANCES MADE

This ad of September 1929 shows the Sunbeam toaster with "Level Turnover Rack" and suggests that one "build sandwiches backwards . . . that is, filling and then toasting them."

• TOASTS TWO SLICES AT ONCE • PUTS WINGS ON YOUR BREAKFAST • HEATS BOTH SIDES AT ONCE • ENTIRELY AUTOMATIC • KEEPS TOAST HOT UNTIL WANTED • REQUIRES NO WATCHING •

New—endowed with the twin gifts of speed and beauty—the Edicraft is the modern toaster. Simply slip in two slices of bread —and close. Both sides of both slices are toasted at once. No watching, no waiting, no burned toast or burned fingers...When the toast is finished, the Edicraft opens up like a morning-glory! Toast is done to just the shade of brown you prefer. You ac-

The Automaticrat of the
BREAKFAST TABLE

tually order the toast by number—setting the lever for the tint you prefer . . . More—the Edicraft will obligingly keep your toast hot until you are ready . . . The Edicraft is made in the Edison Laboratories at Orange, New Jersey, and is the only electric toaster authorized to carry the personal signature of Thomas A. Edison.

*Thomas A. Edison, INC.
ORANGE, NEW JERSEY*

Edicraft *speed* Toaster

The Edicraft Speed Toaster was rather unique. Introduced in 1929, this model "opens like a morning glory" when the toast is done and keeps it hot. If you can find this one, it should be worth **$50.**

wise woman

She knows a gift **MUST** be new and unusual . .

PROCTORS USEFUL, TOO!

Automatic Toaster
Black & Chromium
$5⁹⁵

PROCTOR Toasters and Wafflers are the most appreciated of gifts! They're so useful—and the exclusive **Glow Cone** feature makes them thrillingly new and unusual. Just set the dial to light or dark and the **Glow Cone** silently signals when toast or waffles are done to exact taste. The current goes off and on automatically. The Toaster toasts two slices, keeping them warm until served, without burning. The Waffler even signals when first to pour the batter. Both are beautifully designed. Make Proctors your gift!

Automatic Waffler
Chromium
$9⁹⁵

Proctors are so new and different it will be worth your while to find the store that sells them . . . probably your favorite department store, appliance shop or electric company. Or send order direct to Proctor & Schwartz Electric Co., Dept. 11, 7th and Tabor Road, Phila., Pa.

PROCTOR
heat-controlled
Toasters - Wafflers - Irons

In 1933 Proctor's Toasters and Wafflers featured heat controlled "Glow Cones."

UNIVERSAL

UNIVERSAL ELECTRIC

Toast - double-quick

OVEN TOAST made right at the table, two slices at a time, and with an electric toaster that needs no extra table space.

The Toast rack of this new UNIVERSAL Toaster tilts forward at the touch of a finger. Toast not instantly used may be left on the open toast rack where it will be kept warm and crisp.

"Blue Diamond" chromium plate with black enamel top and base

$7⁹⁵

Ideal for families that demand extra capacity, and ideal too, for families that demand extra quality, for bread placed in the even heat of this toaster's miniature oven is browned quickly and uniformly.

Other Models from $8.95 to $15.95

There is a UNIVERSAL Appliance for every household need—built to uphold a 90-year tradition for Quality —priced to maintain the UNIVERSAL reputation of supreme value and service.

UNIVERSAL ELECTRIC MIXERS
$16.25 to $27.00

UNIVERSAL ELECTRIC SANDWICH TOASTERS
$7.95 to $11.00

UNIVERSAL ELECTRIC HOUSEHOLD IRONS
$2.75 to $6.95

UNIVERSAL ELECTRIC PERCOLATORS
$4.50 to $13.95

LANDERS, FRARY & CLARK
NEW BRITAIN ∴ CONNECTICUT

In March of 1935 Universal boasted toast "double quick" with two slices at a time.

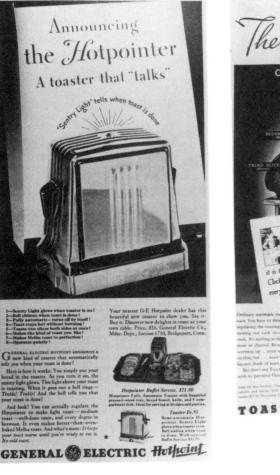

G.E. Hotpoint announced a
"toaster that talks" in 1935.
When the light went out, a
bell would ring ("tinkle, tinkle"). It was sold alone for
$16.95 or with buffet service
which included pressed
wood tray, bread board and
knife, and 5-compartment
dish for $21. Semi-automatic
model complete with service
set sold for $11.95. Current
value for either complete set,
$50.

This Toastmaster ad ap-
peared in a February 1935
issue of *The Saturday Evening
Post.*

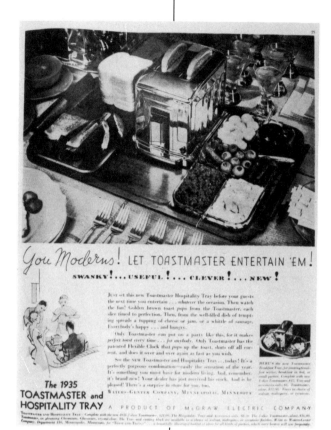

You Moderns! LET TOASTMASTER ENTERTAIN 'EM!

SWANKY! ... USEFUL! ... CLEVER! ... NEW!

The 1935
TOASTMASTER and
HOSPITALITY TRAY

This ad shows the 1935 model "Toastmaster and Hospitality Tray."

UNIVERSAL

UNIVERSAL
AUTOMATIC TOASTER

Delivers toast, automatically,
the exact degree of crispness
you prefer. A demonstration by
your dealer will convince you.

Housekeeping is Easier
and Cheaper when
Electricity
is the Chief Servant
in the Home

LANDERS, FRARY & CLARK
NEW BRITAIN, CONNECTICUT

This universal toaster of 1937 had top slots to accept bread, but when toast was done the side opened up automatically. Current value, **$35.**

THERE'S no age limit to playing the host or hostess with a Toastmaster Toaster. It might have been designed for children, it's so easy to use—and so much fun.

It's *fully automatic.* Just set the adjustment button for light, medium or dark, and that is what you get. The Flexible Timer allows more time when the toaster is cold, less when it's hot. Then, on the split second of perfection, up pop the golden-brown slices, both sides done to a turn, and off goes the current. No watching, no guessing, no burning, no turning!

No burnt fingers. No burnt toast.

No wonder children love to use a Toastmaster Toaster. No wonder husbands grin their appreciation at breakfast time—when a Toastmaster Toaster is on the job.

And no wonder favored ladies consider themselves lucky when some thoughtful friend bestows a Toastmaster Toaster on them as a gracious gift.

On sale, with other fine Toastmaster Prod- ucts, wherever quality appliances are sold.

❧ ❧ ❧

FREE! "The Party's On"—A new and interesting booklet on entertainment ideas and games for young and old. Ask for your copy wherever Toastmaster Products are sold, or write direct to: McGraw Electric Company, Toastmaster Products Division, Dept. 116, Minneapolis, Minn.
European Sales Office: Frank V. Magrini, Ltd., Phoenix House, 19-23 Oxford St., London, W. 1, Eng.

TOASTMASTER *Toaster*

TOASTMASTER PRODUCTS—2-slice fully automatic toaster, $16.00; with choice of Hospitality Trays, $19.95 or $23.50; 1-slice fully automatic toaster, $10.50; Junior toaster, $7.50; automatic Waffle-Baker, $12.50

This 1937 Toastmaster toaster was "fully automatic." Prices ranged from $7.50 for the Junior toaster to $23.50 for a two-slice toaster with deluxe hospitality tray.

THE NEW *Sunbeam* TOASTER

Either "POPS UP" THE TOAST *or* KEEPS IT WARM IN THE TOASTER-OVEN *as you choose!*

How do you like your toast, madam? Like it to "pop up" when it's done . . . or like it to be kept warm in the toaster-oven 'til you're ready to butter and serve it. Whichever way you like, *you get it* in the new Sunbeam Automatic. A touch of the switch sets it to operate either way.

And in either event you simply drop in two slices of bread, press the lever—and

forget it. The Sunbeam toasts faster and its patented Double-Thermostatic control tends to everything. The thermostats respond *fast* when the toaster is cold, and *speed up* as the toaster *warms up* with successive toastings. Every slice the identical shade of the one before it, no matter how many you toast. Also has the hinged crumb tray that drops down for easy cleaning. And *beauty!* The Sunbeam speaks for itself.

STUNNING TRAYSETS

New Sunbeam Toaster Tray set. Lovely circular chrome-plated Tray with satin finish center, rich bakelite handles, and 3-compartment crystal appetizer dish. A stunning combination. Complete set includes Toaster. No. T14.

What could be smarter for buffet luncheons and suppers, spur-of-the-moment evening snacks, all informal entertaining than this lovely Sunbeam buffet set. Has four of the smart new lap-trays of genuine Intaglio Crystal, large 3-compartment appetizer dish. Each piece of glassware has a soft, white, etched center design of unusual distinction. Large, roomy walnut tray. Complete Buffet Set includes Toaster. No. T10.

These two models (this page and opposite) of the Sunbeam "Silent Automatic Toaster" would either pop-up or keep toast warm until time for buttering. Either came alone or with "stunning traysets."

THE FINEST TOASTING TOASTER MADE

If you like delicious, golden-brown toast, buttered *hot* and served *hot*—the way good toast is best—you'll agree with the users of the Sunbeam Silent Automatic . . . that this toaster is "tops."

Not only does it make perfect toast every time—each slice a uniform, *even* brown no matter how many you toast—but it keeps the toast hot until you want it. You simply touch a lever and take it from the toaster-oven when you're ready to butter and serve it. Sunbeam Silent Automatic Toaster No. T7.

STUNNING 8 PIECE BUFFET SET

What could be smarter for buffet luncheons and suppers, spur-of-the-moment evening snacks, all informal entertaining than this lovely Sunbeam buffet set! Four buffet plates of genuine Intaglio crystal, designed exclusively for Sunbeam. Each plate has a soft, white, etched center design of unusual distinction. Two, 2-compartment relish dishes in the same beautiful design. Large, roomy walnut tray with convenient cutting block. No. T8.

THERE'S just no end to the busy day of this *stunning* new De Luxe *Toastmaster* Toast 'n Jam Set. Brighter breakfasts—wholesome "spreads" that bring youngsters racing home from school—pantry raids after the show! But you'll never hear a mur- mer from this automatic pop-up type toaster that never burns toast or fingers. . . . The complete set, with toaster, walnut tray, and color- ful Stangl pottery jam and marma- lade jars—only $17.95, wherever the finest electrical appliances are sold.

TOASTMASTER
DE LUXE *Toast 'n Jam Set*

"TOASTMASTER" is a registered trademark of McGraw Electric Company, Toastmaster Products Division, Elgin, Ill. • Copyright 1939, McGraw Electric Co.

This "Toastmaster Deluxe Toast 'n Jam Set" came with walnut tray and Stangl Pot- tery jam and marmalade jars for just $17.95 in November of 1939.

SEARCH the stores over—and come right back to *this!* Here is the gift with everything—the gift of beauty and distinction; of long, long usefulness; of brighter breakfasts for years to come. A gift which somebody will still be thanking you for, many Merry Christmases from now . . .

It is the superb new *Toastmaster* automatic pop-up type toaster, handsomer than ever in its bril- liant 1940 styling—and, as ever, the very last word in toastmanship! With its famous Flexible Timer, it regulates the toasting to a "T," pops up the piping-hot slices, auto- matically clicks off the current. No watching, no turning, no burning —just perfect toast every time!

See this smartest of toasters ($16.00: 1-slice model, $9.95) and other *Toastmaster* products wherever fine appliances are sold.

"TOASTMASTER" is a registered trademark of McGraw Electric Company, Toastmaster Products Division, Elgin, Illinois • Copyright 1939, McGraw Electric Company

The new 1940 Toastmaster toaster appeared in a Decem- ber 1939 ad for $16. Cur- rent value,
$15 (it hasn't lost much ground!).

Waffle Irons and Sandwich Grills

■

W affle irons and wafer irons date to the fourteenth century and were often used in connection with religious services. Mary Earle Gould described them in *The Early American House* as having "long handles (so the user could stand as far from the heat of the fireplace as possible), with two heads, shutting like pincers. The waffle iron had oblong heads of different sizes, while the wafer iron had round or elliptical heads. The waffle iron, just as today, had a waffle pattern on each head, which gave it its name; later ones made in factories have a number and date of manufacture . . . it was customary to give the bride a wafer iron on which were her initials, the date of the wedding and a hex mark for luck."

The Griswold Manufacturing Company of Erie, Pennsylvania manufactured "on-top-of-the-stove" waffle irons as early as 1865. The American Housewares Institute states that "not until after World War I did the electric waffle iron appear." However, there is a Westinghouse model (a little rectangular device with mechanical lid guaranteed to burn you if not careful) with a date of 1912.

Like the early toasters, the early electric waffle irons worked on the principal of "remove when done to desired crispness." In 1920 Armstrong Electric and Manufacturing Company introduced one with a heat-indicator to show when it was ready to use. Innovations again followed, the twin reversible, the double-decker, and even small dessert-sized waffle makers. Numerous sandwich toasters were designed with reversible or replaceable griddles that became waffle irons.

As manufacturing technology improved, waffle irons and sandwich grills became more sophisticated in operation as well as appearance. The electric waffle iron has changed little over the years. With the exception of Teflon, nothing has drastically changed since the introduction of the little light on top.

9

Waffle Irons

Waffle Iron: 1930s. Berstead Mfg. Co., Fostoria, OH. & Oakville, Ont., Canada.

Low profile rounded chrome body on little curved Bakelite feet and front drop handle. Top heat indicator and wheat shaft decorations. Detachable cord. (Haven't we seen this in several bogus UFO photographs?) **$30**

Waffle Iron: 1920s. Club Electric Waffle Mold, Baltimore & Chicago.

Model #07. Round nickel body on pierced stand base. Has upturned black wooden handles. Slight round elevation on top. 5 tab feet. This one is in somewhat less than desirable condition.
$25 (good condition).

Waffle Iron: Early 1930s. Coleman Lamp & Stove Co., Wichita, KS.

Yes, even Coleman did it and with style! This high Art Deco design has 7½″ round plates that blend into 14″ oval base. Black Bakelite handles. Of note is the small black and white enameled insert on top of leaping gazelle. Sm. lite/dark lever above front handle. Detachable cord. Well done (no pun intended), Coleman! **$45**

Double Waffle Iron: 1940s. Dominion Electric Corp., Mansfield, OH.

Chrome rectangular stepped body with round molds. Top circular decoration. Natural wooden handles and individual light/med./dark wood tipped slide controls. Also has heat light indicators faceted like little jewels. Note special cord with two heads.
$35

Double Waffle Iron: 1940s. Electrahot, Mansfield, OH.

Two 6″ sets of plates mounted on oval base. Tops have heat indicators with surrounding decoration. No controls. Single detachable cord.
$30

Waffle Iron: 1920s. "Star." Fitzgerald Mfg. Co., Turrington, CT.

Measures 7″ and is heavily constructed. Sits on solid flared base. Unique handle design locks in two positions; for raising lid and for carrying. Detachable cord.
$35

Waffle Iron: Early 1940s. General Electric.

8″ round chrome body has ivory Bakelite handles and heat control/off lever. Top decoration is rather patriotic with circle of stars surrounding stripes and leaves. Detachable cord.   **$35**

Waffle Iron: Early 1920s. Hotpoint. Edison Electric Mfg. Co., Ontario, CA.

"Edison" on bottom but proclaims "Hotpoint" in embossed letters on front. Nickel body with angled, faceted top edge and interior dome. Upturned side handles and large front knob in black wood. Stands on pierced base with nice decoration. (Base edge and top decoration match.) Detachable cord.   **$30**

Waffle Iron: 1920s. Hotpoint. Edison Electric Mfg. Co.

Fancy decorated top. $7\frac{1}{2}$″ plates in nickel body and single wooden black knob. Base has metal handles. Note off/on switch on base front. Detachable cord.   **$30**

Waffle Iron/Hot Cake Grill: 1920s. Majestic Electric Appliance Corp., San Francisco, CA.

Innovative with 8″ round reversible plates. Little pierced tower on top has Bakelite cap that serves as a foot for use as double grill (2 hot cakes at one time). Body is of nickel with brown Bakelite swing front handle. **$45**

Waffle Iron: 1920s. Hotpoint (Edison, General Electric) Chicago, IL. Ontario, CA.

"Automatic" below front handle. Turning "cold/hot" in small window on front. Round chrome body with nice top decoration. Ivory Bakelite handles and scalloped base decoration. Detachable cord. **$35**

Waffle Iron: 1930s. Knapp-Monarch.

Small, 6″ round plates. Low profile in nickel body and single black wooden handle. Light Art Deco top decoration. Detachable cord. **$15**

Waffle Iron: 1920s. Majestic Electric Appliance Corp.

This hot cake/waffle maker has top wooden knob and front wooden handle that sticks up far enough to serve as a foot when open for double grilling. Low flared base, reversible plates and detachable cord. **$25**

Waffle Iron: Early 1920s. Manning-Bowman.

7″ round plates in round nickel body with octagonal domed lid. Single wooden handle. Exactly the same as previous waffle iron with exception of base stand which has pierced design. **$25**

Waffle Iron: Early 1920s. Manning-Bowman.

7″ round plates in round nickel body with octagonal domed lid. Single wooden handle. Stands on little cabriole legs with attached tray and detachable cord.   **$25**

Waffle Iron: 1920s. Manning-Bowman, "Homelectrics."

Heat indicator and top perimeter decoration in nickel body with 8″ round plates. Stands on plain stand base and has black wooden handles.   **$25**

Waffle Iron: Mid 1920s. Manning-Bowman, Meriden, CT.

7½″ round nickel body mounted on tall, plain, flared base has large, single black wooden front knob. No decoration. Flat prongs for detachable cord.   **$15**

Waffle Iron: Late 1930s. "Twin-O-Matic," Manning-Bowman.

Fabulous design, like a work of art. Heat indicator on top has rotating knob from Off to Dark in numbered increments. Whole body flips over on brown Bakelite stand mounted on chrome base. Was also available in non-automatic version. **$75**

Waffle Iron: Late 1930s. Manning-Bowman.

Isn't this a pretty one? Part of a set, has 7″ round plates in chrome on flat, square base. Black Bakelite handles have reeded line designs that also appear on borders. Heat indicator on top. Has original paper booklet "It's Waffle Time." High Art Deco design. **$40**

Waffle Iron: Early 1940s. Manning-Bowman.

Double model with parallel stripes on either side of dual heat indicators. Drop swing walnut handles, attached base and brown Bakelite feet. **$20**

Waffle Iron: 1920s. "Simplex," S.H. Co.

This menacing device is made of cast iron and measures $10\frac{3}{4}'' \times 12\frac{1}{4}''$ and weighs a whopping 20 lbs. Iron waffle plates are housed in a well and are hinged. One wooden handle operates movement of plates. Note open electrical connectors that are only protected by a metal surrounding plate at rear. Plates lift out for cleaning. Could possibly be a commercial product. **$40**

Waffle Iron: 1920s. "Torrid," Beardsley & Wolcott Mfg. Co., Waterbury, CT.

Good basic design in nickel with $7\frac{1}{2}''$ plates highlighted by green handles and knob as well as green tab feet. Window on front indicates "too cold" or "too hot" and "bake." Detachable cord. **$35**

Waffle Iron: 1910s. Universal, Landers, Frary & Clark, New Britain, Ct.

E 931. $7\frac{1}{4}'' \times 4''$ rectangular nickel body on slightly angled legs and black wooden bun feet. Single upturned front handle. Requires double-headed plug.
$\qquad$ **$25** (with cord).

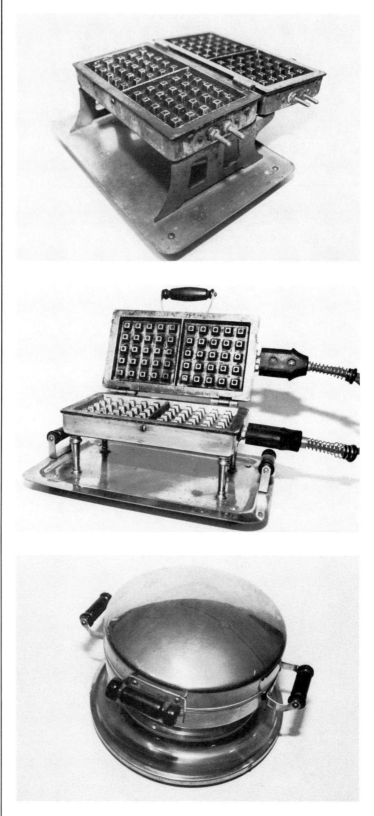

Waffle Iron: 1910s. Universal, Landers, Frary & Clark.

E 9300. $7\frac{1}{4}'' \times 4''$ rectangular nickel body on flat end, pierced legs that extend to form rest for top plate. Basically same as E 931 but also has attached tray and requires double-headed plug.
$30 (with cord).

Waffle Iron: 1910s. Universal, Landers, Frary & Clark.

E 9305. $7\frac{1}{4}'' \times 4''$ rectangular nickel body on round legs with attached tray. Again basically identical to previous models with different legs.
$35 (with cord).

Waffle Iron: Early 1920s. Universal, Landers, Frary & Clark.

Beautiful and plain. Basic clean design with $7\frac{1}{2}''$ round plates on solid flared base. Turned black wooden handles, detachable cord. **$25**

Waffle Iron: Early 1920s. Universal, Landers, Frary & Clark.

$7\frac{1}{2}''$ round plates. Nickel body has incised parallel line decoration that also appears at perimeter of base. Fancy mounted drop front handle and straight-up mounted base handles. Detachable cord.   **$20**

Waffle Iron: Early 1920s. Universal, Landers, Frary & Clark.

"The Trade Mark Known to Everyone." 8″ round plates and domed lid. Upturned black wooden handles. Stands on pierced, flared base. Detachable cord. **$20**

Waffle Iron: 1930s. Universal, Landers, Frary & Clark.

Part of a large breakfast set. Floral decorated porcelain set into chrome body on pierced, round pedestal base with upturned handles. Fancy mounted drop wooden handle and light/dark adjustment. Painted wooden knob feet.
          $60

Waffle Iron/Sandwich Grill: 1930s. Universal, Landers, Frary & Clark.

Rectangular rounded body with interchangeable plates, mounted base, detached cord. Front swivel handle serves as a foot for double grilling. Wheat decoration on top.

$25 (with 2 sets of plates).

Unmarked Waffle Iron: Late 1920s.

The design is great. Light weight construction, low profile, upturned open handles, leaf top decoration. Body and base have border of palmettoes. Leafy drop, bail handle and front heat indicator. Detachable cord.

$35

Unmarked Waffle Iron: 1920s.

Once marked but not anymore! To make it worse, it was our fault. In trying to clean the front plate, the whole thing disappeared. Plain round nickel body with 7″ plates and upturned wooden handles. Stands on some of the ugliest legs we've seen and seems to tiptoe on little tab feet. **$15**

Unmarked Waffle Iron/Sandwich Grill: 1930s. ("N" in circle crossed by lightning).

Two slice sandwich grill with aluminum plates, front drip spout. Plates not reversible but changeable. Rectangular nickel body attached to base. Heat indicator on top and line decoration. Hinges open wide to accept thick sandwiches. Detachable cord.

$25 (with both sets of plates).

Unmarked Waffle Iron: 1930s.

Plates are light weight aluminum, measure 5″ square, and are set into nickel body on angled legs and little pad feet. Green or red plain wooden handle, attached cord. Inexpensively made but cute. Waffle plates have rounded humps. Would be perfect for a quick cheese sandwich. **$18**

Unmarked Waffle Iron: 1930s. (Manufacturer plaque missing). Possibly Universal or Royal Rochester.)

Nickel body with bird and floral transfer on porcelain inset. Green painted turned wooden handles. 5 tab feet with one directly below hinge help support weight. Without front manufacturer plate no clue is available to origin except shape and style. **$50**

Waffle Iron: Pat. Dates
1905–1921. Westinghouse
E. & M. Co., East Pitts-
burgh, PA.

One of Westinghouse's earliest waffle irons. Rectangular
chrome body and mechani-
cal handle with wooden
hand hold. Removable ca-
briole legs slip into bottom
slots. Attached cord with
off/on switch. Handle folds
forward—could easily burn
one's hand.   **$75**

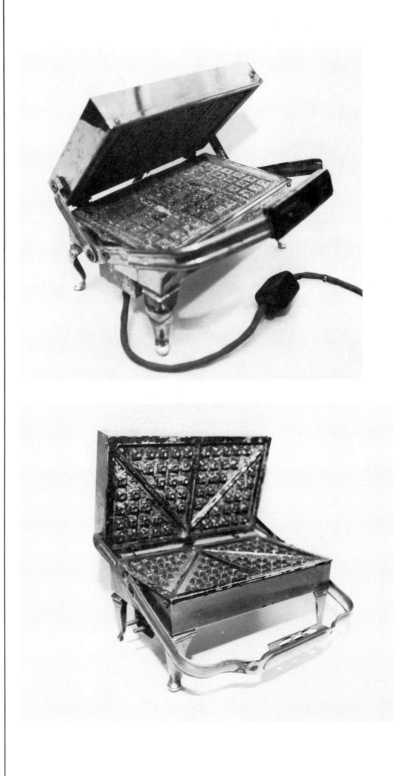

Waffle Iron: 1920s. West-
inghouse E. & M. Co.,
Mansfield Works, Mans-
field, OH.

Very similar to other early
Westinghouse waffle makers.
Removable cord, straight, removable legs and all metal
mechanical handle. Nickel
rectangular body. Guaran-
teed to burn you when
opened.   **$35**

Sandwich Grills

Sandwich Grill: 1930s. "Hostess." All Rite Co., Rushville, IN.

5″ square nickel body with black screw-off handle. Stands on angled legs made into lower body. This has original box and paper "suggestions" pamphlet. Side of box declares "For Toasting plain & 3-decker sandwiches, toasts dry, buttered, wafer thin Melba toast from top to bottom in 20 seconds". **$40**

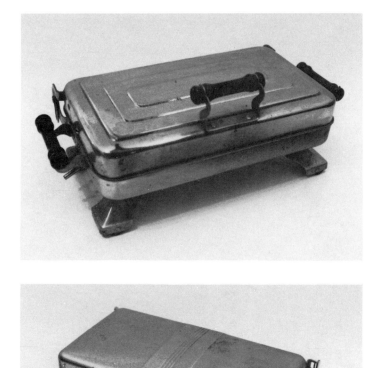

Sandwich Grill: 1920s. "Victorian." Berstead Mfg. Co., Fostoria, OH.

Don't know how they got the name for this early Art Deco design! Rectangular nickel body 10″ × 5½″. Permanent plates and flared legs. Black wooden handles mounted on curved mounts. Detachable cord. **$20**

Sandwich Grill: 1930s. "Eletrex" (Partial paper label on base reads "sold only at Rexall Drug Stores").

Rectangular nickel and black painted metal base and tray. Black wooden handle. Plain design. Inexpensively constructed. Detachable cord. **$12**

Sandwich Grill: 1930s.
"Lady Hibbard." Hibbard,
Spencer, Bartlett & Co.,
Chicago, IL.

Nickel rectangular body on
cabriole legs. Black wooden
side handles. Note front
handle that swivels to form a
"foot" for top plate enabling
use of both plates as grill.
Drip spout in front. Top has
Art Deco design.   **$18**

Sandwich Grill/Waffle
Iron/Frying Pan: 1920s.
Manning-Bowman.

Interchangeable plates in
rectangular nickel body $10\frac{1}{2}''$
$\times$ $6\frac{1}{2}''$ with drip tube. In-
dented grill plates serve as
frying pan. Black wooden
handle and attached cord.
  $35 (with 2 sets of plates).

Sandwich Grill: 1930s. Porcelier Mfg. Co., Greensburg,
PA.

Part of a large breakfast set.
All Porcelier electrical appliances completely enclosed
by porcelain bodies. This
sandwich grill with basket
weave top decoration also
has multi-colored floral
transfer on top. Handles are
one piece with body and
have silver colored line decoration. Plates are non-re-
versible.   **$60**

Sandwich Grill: Early 1940s. Universal, Landers, Frary & Clark.

W 8960. Double-size sandwich grill with drip spout in rectangular chrome body, walnut handles and circular decoration on lid. Detachable cord. **$20**

Unmarked Sandwich Grill: 1920s.

6″ × 14½″ nickel body on short legs. Will accommodate three sandwiches simultaneously. No clue to manufacturer. **$20**

Electric Breakfast Grid

ONLY $6.50

For the price, one could hardly imagine a more useful or delightful Holiday Gift

Discover This Charming New Way to Cook Piping Hot Breakfasts At the Table!

GRIDDLE cakes . . . more griddle cakes . . . each new helping golden brown and steaming hot! French toast, too, bacon, sausage, ham, Breakfast Grid cooks them all without grease or smoke. No wonder women call it a discovery! Foods served from the Breakfast Grid are tastier, hotter. Breakfasts are pleasanter and quicker when cooked this nice new way at the table. And Breakfast Grid leaves you no grease-caked griddle or frying pan to wash afterwards. No electric appliance has yet been developed which excels the White Cross Breakfast Grid in all around usefulness and convenience.

of pure seamless, thick aluminum. With a base of sparkling nickel plate, Breakfast Grid is decidedly impressive in appearance and is fully guaranteed. Standard cord and 2-piece plug included. Operates on either direct or alternating current, 110 to 125 volts, and it costs no more for electric current than the ordinary iron or toaster.

Sent on Six Days' Approval

Every home should have this amazingly efficient servant. Examine White Cross Breakfast Grid at the nearby dealer, or if this is not possible, mail coupon, and we will send one direct to your home on approval. If you are not delighted, at the end of six days, merely slip it into its original package and return it. Your money will be refunded at once. What else can you buy for $6.50 that will bring you such life-long pleasure and convenience.

TOASTER $4.50

Will Last a Lifetime

Ten inches in diameter, exclusive of handles, and made

NATIONAL STAMPING AND ELECTRIC WORKS
3212-50 W. Lake Street, Dept. 2W, Chicago, Ill.

WHITE :: CROSS

DEALERS AND JOBBERS:

In the late 1920s, White Cross introduced their "Electric Breakfast Grid" for griddle cakes, French toast, bacon, sausage & ham. They would even send one to you "on approval."

★ Hotpoint Electric Gifts
Bring Lifetime Appreciation

They may cost even less than you have planned to spend. Yet these beautiful gifts, for all the years to come, will be daily reminders of your thoughtfulness and affection. Hotpoint quality will endure.

The Gifts Shown Above

New Hotpoint Percolator Set — Makes delicious coffee in 10 minutes with no attention. 6-cup percolator set, nickel finish, large tray, gold lined sugar and creamer. $17.00.

Hotpoint Toast-Over Toaster — Makes perfect toast quickly right at the table. When one side is toasted, simply turn the ebonized knob and the toast turns over automatically. $8.00.

3-lb. Traveling Iron Set — Ideal for the girl at school, to press or iron things in a jiffy. Ivory handle; pearl silk cord. Handsome, permanent gift case. $7.50.

There's a Hotpoint dealer near your home. You will know him by his "Hotpoint Servants" sign—or by the Hotpoint appliances and cards in his window. See his full line of beautiful, economical Hotpoint gifts. Insist on "Hotpoint" electric appliances. For over 20 years they have given perfect satisfaction.

EDISON ELECTRIC APPLIANCE CO., Inc.
Chicago • Boston • New York
Atlanta • Cleveland • St. Louis
Ontario, Calif. • Los Angeles
San Francisco • Portland • Seattle
Salt Lake City
In Canada:
Canadian General Electric Co., Ltd.
Toronto

Hotpoint "Dolly Madison" Percolator Set — The gift de luxe. Decoration and lines are of the Adam period. Its beauty and utility will bring happiness for a lifetime. Pieces may be purchased singly from time to time if desired. Set complete: Satin Silver plate, $65.00. Polished nickel, $55.00.

Hotpoint Waffle Iron — What could be more appreciated! Makes delicious, golden brown waffles right at the table. Saves steps. No smoky fumes. Bakes both sides at once without turning. No greasing necessary. Recipes packed in the box with it. Highly nickeled to STAY beautiful. With cord and plug $15.00.

Hotpoint
SERVANTS

WORLD'S LARGEST MANUFACTURER OF HOUSEHOLD ELECTRIC HEATING APPLIANCE

Irons   Electric Ranges   Curling Iron   Ironing Heaters

Christmas ad of 1925 shows Hotpoint gift ideas and features the "Dolly Madison" percolator set and waffle iron.

$9.00
In Canada $11.00

A tang in the frosty air—in the house a most appetizing aroma—waffles crisping—big brown waffles piping hot from the STAR-Rite Waffle Iron—delicious!

And that's just one of the good things cooked quickly to perfection right at the table on the STAR-Rite Waffle Iron—tempting omelets, cookies, short cake, pan cakes—wonderful goodies with a flavor that brings folks back for more.

Splendidly designed, the STAR-Rite Waffle Iron is a fine table appointment—all metal parts except the grids are of brass, finished in sparkling nickel—it's RUST PROOF.

And it's most remarkably efficient, too—there is a heating element in both the upper and the lower grid, and these pure aluminum grids are die cut, they do not stick—no greasing is needed.

Equipped with special non-heating carrying handles, non-heating lid lifter, cord, plug and switch—complete at $9.00. You'll enjoy the STAR-Rite recipes—send the coupon today.

STAR-Rite Reversible Toaster
$5.00
In Canada, $6.75
Toasts two large slices at once—most convenient, non-heating turning handles nickel finish, splendidly made.

STAR-Rite
ELECTRICAL NECESSITIES

FITZGERALD MANUFACTURING COMPANY
Torrington   Connecticut
CANADIAN FITZGERALD COMPANY
95 King Street East,   Toronto, Ontario

COUPON
FITZGERALD MANUFACTURING COMPANY
Torrington, Conn.
Please send me the STAR-Rite Grill Book, describing the latest member of the STAR-Rite family, including recipes.
Name
Address
City

This Star-Rite waffle iron pictured in a late 1925 ad featured a handle that would lock in a lifting or carrying position.

In 1925 this Sun Chief Electric sandwich toaster came in nickel or chrome. Note the unique design of the handle. **$35** (current value).

159

Star-Rite "Electrical Necessities" thought everyone should come indoors and eat waffles after outdoor sports in this November 1927 *Good Housekeeping* ad.

The manufacturer of this table cooker, available with griddle or waffle surfaces, was not given in a *Good Housekeeping* article of July 1929. It was similar to many others made at the time.

52 Treats
made with THE HOTPOINT WAFFLE IRON!

The Hotpoint Waffle Iron illustrated has a convenient "On" and "Off" toggle switch in the base. Price $9.45. Other models up to $18.50. All have the practically indestructible Hotpoint CALROD element. They bake evenly, both sides at once, without "sticking" and without grease, smoke or odor.

H OTPOINT WAFFLE IRONS are not just "waffle" bakers. Of course they do bake most delicious, golden brown waffles that melt in your mouth.

But, in addition, any *Hotpoint* Waffle Iron will make 52 varieties of good things to eat.

A different delight for each Sunday night

Here are some of the things which you can bake right at the table: chocolate cookies, jelly sandwiches, shortcake, apple fritters, cheese sandwiches, corn bread, sponge cake, cocoanut delicacies. These and 44 other recipes, 52 in all, are given in the helpful recipe book that comes *free* with every Hotpoint Waffle Iron. This convenient recipe book also gives delightful menu suggestions for breakfasts, luncheons, bridge parties and dinners. With its help you'll find a Hotpoint Electric Waffle Iron one of your most treasured, resourceful aids for entertaining.

Ask your electric company or dealer to show you the several styles of Hotpoint Waffle Irons. Whichever model you select, you'll find your copy of the recipe book inside the carton. (If you already have a Hotpoint Waffle Iron we will gladly send the new recipe book on receipt of ten cents.)

Hotpoint ·

EDISON ELECTRIC APPLIANCE CO., Inc.
A GENERAL ELECTRIC ORGANIZATION
5600 West Taylor Street   Factories: Chicago, Ill.
Chicago, Illinois   and Ontario, Calif.
© 1929 E. E. A. Co., Inc.

World's Largest Manufacturer of Household Electric Heating Appliances and Electric Ranges

Hotpoint waffle makers of 1929 had an off/on toggle switch and included a recipe book with 52 varieties of waffles.

Maryland Cream Waffles are light and crisp. Their easy recipe accompanies the Manning-Bowman Waffle Iron, 1621, $15.

A *Good Housekeeping* article in 1929 featured this Manning-Bowman waffle maker.

The Westinghouse waffle iron of 1929 pictured in this ad featured a new "Watchman" heat control device that told if the iron was too hot or too cold. It kept the heat automatically adjusted, turning itself on or off as necessary.

Now! Virginia Sweet Pancakes made at the table!

Virginia Sweet Electric Griddle
with Automatic Heat Control

Postpaid for two package tops and $**2**.95

Tested and approved by
Good Housekeeping Institute

COME out of the kitchen! Bake your pancakes sociably at the breakfast table, where you can enjoy your menfolk's company while they enjoy your handiwork!

You'll be proud of this modern griddle on your table. Its polished nickel and heavy aluminum, its red handles and red metal tray reflect the gayety of Virginia Sweet breakfasts.

And how you will thrill to the marvelous taste of Virginia Sweet Pancakes baked on this greaseless griddle — the true delicate flavor of wheat, corn, rice — the 3 Staffs of Life.

Not a chance of failure. You just add milk or water to Virginia Sweet, and the griddle's automatic heat control assures perfect baking. 10″ diameter. Bakes 3 cakes at a time.

Designed especially and exclusively for Virginia Sweet Pancake Flour or Buckwheat Flour. Tested and approved by

Good Housekeeping Institute. Guaranteed against defects for one year. Lava insulation insures many years of continuous use.

To Obtain This Griddle

Hardware dealers say that they easily could get $7.50 for this Virginia Sweet Griddle. But we will send it direct to you postpaid for $2.95 and the tops from two standard or one large package of Virginia Sweet Pancake Flour or Buckwheat Flour.

If your dealer can't supply you, send us $3.25 and we will mail you, postpaid, the griddle and two full-sized packages of Virginia Sweet. Add 30c for shipments to Rocky Mountain States and west; also Canada.

THE FISHBACK COMPANY, Indianapolis, Indiana
Manufacturers of Nationally Advertised Food Products

VIRGINIA-SWEET
PANCAKE FLOUR
BUCKWHEAT FLOUR and SYRUP

Virginia Sweet Pancake Flour advertised this premium, "The Virginia Sweet Electric Griddle," in 1929 for $2.95 and two box tops. (Don't know who manufactured this for V. Sweet.)
$20 (good condition).

230

This $7.50 Electric Pancake Griddle

(*with Automatic Heat Control*)

only . . . $2.95

with 2 box tops from Virginia Sweet Pancake Flour or Buckwheat Flour.

TWINKLING polished nickel and heavy aluminum on a red metal tray, this amazing new table griddle transforms the baking of pancakes into a sociable dining room art.

Not a chance of failure. You just add milk or water to Virginia Sweet Pancake Flour, and the griddle's automatic heat control assures perfect baking. 10″ diameter. Bakes 3 cakes at a time.

And oh, the flavor of Virginia Sweet Pancakes baked on this Virginia Sweet Griddle — the true delicate taste of wheat, corn, rice—the 3 Staffs of Life—in the Virginia Sweet blend.

Designed especially for Virginia Sweet Pancake Flour or Buckwheat Flour. Guaranteed against defects for one year. Lava insulation insures many years of continuous use.

To Obtain This Griddle

Send us $2.95 and the tops from two standard or one large package of Virginia Sweet Pancake Flour or Buckwheat Flour. If your dealer can't supply you, send us $3.25 and we will mail you, postpaid, the griddle and two full-sized packages of Virginia Sweet. Add 30c in the Rocky Mountain States and west; also Canada.

THE FISHBACK CO.
Indianapolis, Ind.
Manufacturers of Nationally Advertised Food Products

Tested and Approved 2856
Good Housekeeping · Institute
GOOD HOUSEKEEPING MAGAZINE

VIRGINIA-SWEET
PANCAKE FLOUR
BUCKWHEAT FLOUR *and* SYRUP

Another ad for the "Virginia Sweet Pancake Griddle" appeared in November 1929.

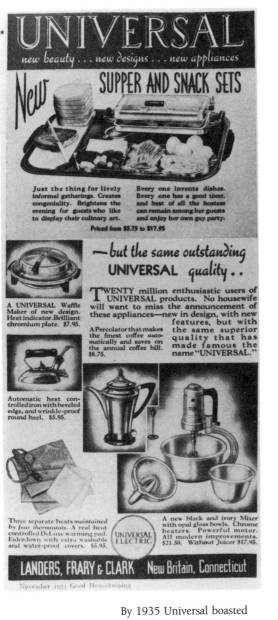

FREE! "The Party's On"—A new and interesting booklet on entertainment ideas and games for young and old. Ask for your copy wherever Toastmaster Products are sold, or write direct to: McGraw Electric Company, Toastmaster Products Division, Dept. 139, Minneapolis.

Toastmaster WAFFLE BAKER

TOASTMASTER PRODUCTS—Automatic Waffle-Baker, $12.50; 2-slice fully automatic toaster, $16.00; with choice of Hospitality Trays, $19.95 or $23.50; 1-slice fully automatic toaster, $10.50; Junior toaster, $7.50

A December 1937 ad showing the "Toastmaster" automatic waffle baker.

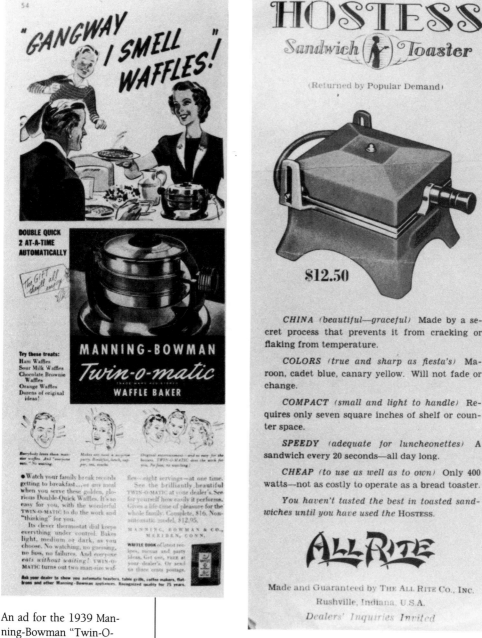

An ad for the 1939 Manning-Bowman "Twin-O-Matic" waffle baker. An outstanding design in chrome and brown Bakelite.

The Hostess Sandwich Toaster by the All Rite Co. was reintroduced with china clad body that came in colors "true and sharp as fiesta's."

Appendix

∎

Manufacturers

This Appendix, designed as a quick alphabetical reference of small appliance manufacturers in the United States, will give you a good overall picture of the industry as a whole. Many companies have come and gone and not much is known about some of the smaller ones. There have been mergers and take-overs, especially among the larger manufacturers. Many started out in entirely unrelated fields. This chapter will cover most of the manufacturers in the book, giving location, what they manufactured, and any known dates.

All Rite Co.
Rushville, Indiana. Manufacturer of the "Hostess" sandwich toaster. In operation from the 1920s–1930s.

Angelus
A Division of The Simplex Corporation. Manufacturer of the Campfire Bar-B-Q marshmallow toaster which appeared in the 1906 Simplex catalog.

Armstrong Mfg. Co.
Originally the Armstrong Standard Stamping Company of Marysville, Ohio, and later of Huntington, West Virginia. In 1916 introduced an ingenious table stove ("the stove that cooks three things at once") and manufactured the "Perc-O-Toaster" (1918) and waffle irons. Still in operation although no longer manufacturing appliances.

Brannon, Inc.
Detroit, Michigan. Manufacturer of the "Cord-Less-Matic" iron, patent pending in the 1930s.

Calkins Appliance Co.
Niles, Michigan. Manufacturer of the "Breakfaster" (combination toaster and hotplate) in the 1930s.

Chronmaster
New York and Chicago. Manufacturer of "Mixall" drink mixer, patent date 1934.

Coleman Lamp and Stove Co.

Wichita, Kansas. Yes, even Coleman made a waffle iron in the 1920s and 1930s. Currently manufactures lanterns and camping equipment.

Dominion

Early manufacturer of electric popcorn poppers and later full range of appliances. Bought by Scovil circa 1969.

Dover Mfg. Co.

Dover, Ohio. Manufacturer of a child's iron between 1919 and 1927.

Mary Dunbar (Chicago Electric Mfg. Co.)

Manufacturers of "Handymix" mixer in the late 1930s.

Edison Electric Appliance Co.

Founded by Thomas Edison. Manufacturer of full range of small appliances. Merged with Thompson-Huston Co. in 1892 to become General Electric.

Estate Stove Co.

Hamilton, Ohio. Manufacturer of clever four sided toaster (pat. 1923) which flipped slices simultaneously.

Eureka Vacuum Cleaner Co.

Detroit, Michigan. Only known cooking appliance of this well-known company was a portable electric range (1930s) which was a small oven with flip-down hotplates.

Everhot

Early manufacturer of slow cookers (pat. 1925). Acquired in 1949 by McGraw Edison.

Farberware (S.W. Farber)

New York. Manufacturer of "Coffee Robot" (1930) with new replaceable fuse and the "Broiler Robot" (1937). Bought in 1966 by LCA Corporation.

General Electric (see Edison Electric Appliance Co.).

General Mills (Betty Crocker)

From early 1940s manufactured a line of appliances including waffle irons, cookers, deep fryers, and mixers. Became part of McGraw Edison in 1954.

A.C. Gilbert

New Haven, Connecticut. Manufacturer of the famous Erector Sets and the "Polar Cub" portable mixer (1929) made specifically as a premium for Wesson Oil Snowdrift. Also made a malt mixer in the 1930s.

Hamilton Beach

Racine, Wisconsin. Founded by Osius, Hamilton and Beach in 1910. Perfected the Universal home motor. Manufacturer of malt machines and mixers. Sold to Scovil 1920–1924.

Hankscraft
> Madison, Wisconsin. Manufacturer of automatic egg cookers and electric casseroles in the 1930s.

Hobart (see KitchenAid).

Hotpoint (see Edison Electric Appliance Co.).

Kenmore (Sears)
> Chicago, Illinois. Manufacturer of mixers and toasters (today major appliances).

KitchenAid (Hobart)
> Troy, Ohio. Manufacturer of the first home coffee mill (1937) and the famous mixer, the first to be mounted on a base (1920).

Knapp-Monarch
> Belleville, Illinois. Manufacturer of a full line of appliances (1928). Bought by Hoover in 1939.

Manning-Bowman
> Cromwell, Connecticut. Founded 1849. First manufacturer of Britiania ware and later coffee makers. After turn of the century entered the electric appliance field with percolators, waffle irons, etc. Sold to Berstead after WW II. Now McGraw-Edison.

Mirro (Mirro Aluminum)
> Manitowoc, Wisconsin. Founded 1908. Manufactured coffee makers and tea kettles in the 1920s. In 1957 changed name to Aluminum Goods Mfg. Co. and is world's largest manufacturer of aluminum cooking utensils.

Mixmaster (see Sunbeam).

Montgomery Ward & Co.
> Many small appliances were made for Montgomery Ward & Co. by various manufacturers from the turn of the century until present.

Porcelier Mfg. Co.
> Greensburg, Pennsylvania. Manufacturer of porcelain body breakfast appliances including toasters (1930s–1940s).

Royal Rochester (Robeson)
> Rochester, New York. Known primarily in the 1920s and 1930s for beautifully decorated porcelain percolators.

Silex (Proctor Silex)
> Philadelphia. Originally Liberty Gauge & Instrument Company (1929). Manufacturer of coffee makers, irons, ice cream freezers, coffee grinders and meat choppers. Still in business as a part of S.C.M.

Sunbeam (Chicago Flexible Shaft Co.)
> Manufacturer of a wide range of electrical products.

Thought to be first manufacturer of first flat toaster under Flexible Shaft Co. name in 1920s. Also made the famous Mixmaster mixers.

Toastmaster (Waters Genter Co.)

Minneapolis, Minnesota. Manufacturer of the first automatic pop-up toaster, 1927.

Universal (Landers, Frary & Clark)

Founded 1857. Became Universal in the 1890s. In 1908 introduced its first percolator. Probably most diversified manufacturer of appliances in U.S. history. Later part of General Electric.

Westinghouse

Mansfield, Ohio. Early and nearly first manufacturer of generators and motors. Manufactured first electric frying pan (1911) and first electric waffle iron found to date (1905). Also made toasters and irons.

Chronology

This list of "firsts" is not meant to be encyclopedic nor exhaustive. It is merely a quick reference for collectors and novices, providing some surprises for even veteran collectors.

1882	Electric generating plant (Thomas Edison, N.Y.). Patent for an electric iron to H.W. Seeley.
1893	All-electric kitchen of the future seen at Chicago World's Fair.
1903	Iron with detachable cord (General Electric). Patent for electric iron with "hotpoint" (H.W. Seeley).
1905	Electric toaster (General Electric, Model X-2). Toaster Stove and electric waffle iron (Westinghouse).
1908	Patents for electric percolator and cold water pump (Landers, Frary & Clark).
1909	Travel iron (General Electric).
1911	Electric frying pan (Westinghouse).
1917	Table stove (Armstrong).
1918	Combination toaster/percolator (Armstrong "Perc-O-Toaster").
1920	Heat indicator on a waffle iron (Armstrong). Mixer mounted on permanent base (Hobart KitchenAid). Electric egg cooker (Hankscraft). Flip-flop toaster (everyone).
1922	Cordless iron (Nocord Electric Co.). Four slice toaster (Estate Stove Co.).
1923	Portable or mounted mixer ("Whip All" Air-O-Mix).

1924	"Automatic" iron (Westinghouse). Home malt mixer (Hamilton Beach #1). Combination flat toaster/grill (Sunbeam). Automatic toaster (D.A. Rogers, not successful).
1926	Automatic pop-up toaster (Toastmaster #1-A-1). Steam iron (Eldec).
1927	Adjustable temperature iron (Liberty Gauge Co.).
1929	Electric iron with "Button Nooks" (General Electric).
1930	Percolator with replaceable fuse (Farberware). Lightweight travel iron under 6 lbs. (Sunbeam).
1933	Toaster to accommodate rolls (Samson United Corp.).
1934	Iron with "Snap Stand" (Proctor, Schwartz Co.).
1937	Home coffee mill (Hobart KitchenAid). Automatic coffee maker "Coffee Robot" (Farberware). Conveyance device toaster ("Toast-O-Later").
1939	Steam iron with Underwriter's approval ("Steam-O-Matic").
1940	Toaster with "keep warm" feature (General Electric).

The first successfully marketed electric toaster. This General Electric model D-12 was introduced in 1906, three years later than the experimental X-2.

The first "automatic" pop-up toaster was the 1926 Toastmaster model 1-A-1. At that time the company name was the Waters Genter Co.

The first electric home coffee grinder was the KitchenAid model A-0 which appeared in 1937. The black ring in the center adjusts the fineness of the grind.

Bibliography

Celehar, Jane. *Kitchens and Gadgets*. Radnor, Pa.: Wallace-Homestead Book Company, 1982.

———. *Kitchens and Kitchenware*. Radnor, Pa.: Wallace-Homestead Book Company, 1985.

Franklin, Linda Campbell. *From Hearth to Cookstove*. 2d ed. Orlando: House of Collectibles, 1978.

Fredgant, Don. *Electrical Collectibles (Relics of the Electrical Age)*. San Luis Obispo: Padre Productions, 1981.

Lifshey, Earl. *The Housewares Story*. Chicago: National Housewares Manufacturers Association, 1973.

Rinker, Harry, ed. *Warman's Americana & Collectibles*. 4th ed. Radnor, Pa.: Wallace-Homestead Book Company, 1990.

Sparke, Penny. *Electrical Appliances (Twentieth Century Design)*. New York: E.P. Dutton, 1987.

Other Helpful Publications

Better Homes & Gardens (monthly issues: 1937–1939).

Good Housekeeping (monthly issues: 1914–1918, 1920–1941).

The Ladies' Home Journal (monthly issues: 1925–1943).

The Saturday Evening Post (monthly issues: 1935).

Woman's Home Companion (monthly issues: 1934–1935).

Here's the New Easy Way to Cook with Delicious Results. New York: A.S. Farber, Inc.

The World's Fair Souvenir Album. Chicago: C. Rapp & Sons, 1894.

Twin Waffle Baker Instructions. Meriden, Ct.: Manning-Bowman & Co.

Kitchen Tested Recipes Using the Famous Sunbeam Mixmaster. Chicago: Sunbeam.

How to Get the Most Out of Your Sunbeam Mixmaster. Chicago: Sunbeam.

Index